Working in Steel

Working in Steel:
The Early Years in Canada,
1883-1935

Craig Heron

M&S

Canadian Cataloguing in Publication Data

Heron, Craig.
  Working in steel

(The Canadian social history series)
Includes bibliographical references and index.
ISBN 0-7710-4086-5

1. Iron and steel workers – Canada – History.
2. Steel industry and trade – Canada – History.
I. Title. II. Series.

HD8039.I52C35 1988   331.7′669142′0971   C87-093714-6

This book has been published with the help of a grant from the Social Science Federation of Canada, using funds provided by the Social Sciences and Humanities Research Council of Canada.

Printed and bound in Canada

McClelland and Stewart
The Canadian Publishers
481 University Avenue
Toronto, Ontario
M5G 2E9

Contents

Acknowledgements

This book would not have been possible without the kind assistance of staff at the Public Archives of Canada, the Public Archives of Nova Scotia, the Ontario Archives, the Beaton Institute of the College of Cape Breton, the Hamilton Public Library, the Sault Ste. Marie Museum and Public Library, the Metropolitan Toronto Central Reference Library, the Dalhousie University Archives and Killam Library, and the libraries of the University of Toronto and York University. It began as a post-doctoral project with the financial support of the Social Sciences and Humanities Research Council and was later helped along with a Labour Canada University Research Grant. Many ideas presented here were refined through presentation to a string of job interviews in the early 1980s. I also got helpful criticism from the Faculty Research Seminar of the York University History Department. The most important influence on the book's evolution was the rich process of discussion in the Labour Studies Research Group in Toronto over the past five years. I am also grateful to Greg Kealey for patience, support, and sound advice, to David Sobel for exemplary work as a perceptive, sharp-eyed research assistant, and to Joann Trypuc for tolerating my obsessiveness in the final stages of writing. Most of all, I am deeply indebted to Robert Storey, who constantly challenged and sustained me with the best kind of intellectual comradeship.

For R.H.S.

Introduction

This is a book about work. It is an attempt to show how, around the turn of the twentieth century, something fundamental began to change in the daily lives of people who worked for wages in Canadian factories. For the men and women who punched the clock at the beginning of ten- to twelve-hour working days, this was not the age of Laurier, Borden, or King, still less the "Edwardian" era. For them this was the age of mass production.[1]

Many of us have glimpsed this new work world through the eyes of Upton Sinclair or Charlie Chaplin, but most Canadian history books are silent about this profoundly disruptive transformation in Canadian society. Several historians have drawn our attention to the dismal standard of living of most urban workers in this period of massive economic expansion,[2] but the conditions of employment underlying this material life have generally remained unexplored. The analysis in this book assumes that a crucially important experience for people who depend on wages for their survival is the nature of their relationship with their employers. In the early twentieth century, a new kind of employer was emerging. More and more workers were facing not individual entrepreneurs who headed up local businesses, but rather powerful new corporations with unprecedented resources to control the productive capacity of the country. The great changes in the world of work in this period flowed in large part from the new corporate strategies for capital accumulation.[3] The throngs of workers who entered through the factory gates found much larger, more impersonal workplaces, confronted much more complex machinery and harsh, autocratic supervision, and, above all, felt increasing pressure to produce more and more, faster and faster. In such settings, they had difficulty establishing much countervailing power for asserting their own concerns. But they never stopped trying.

9

The steelworkers discussed in these pages stand in for a wide range of workers in the new mass-production industries – auto- workers, chemical workers, pulp and paper workers, packinghouse workers, and many others in industries using sophisticated machin- ery in immense factories to produce large quantities of standardized goods. They are not completely "typical," since each of these new workplaces had a rhythm of its own – each with a special combina- tion of markets, management, technology, and patterns of worker resistance and accommodation. In particular, steelworkers were not women, who had to cope with the patriarchal ideologies and structures that shaped their role in the paid work force. Neverthe- less, working in a steel plant was enough like working in other heavy mass-production plants that steelworkers' experience can suggest some of the broad themes of the new twentieth-century work world in Canada's industrial centres.

Over the years the importance of the steel industry and the drama of its work processes have inspired many writers to put pen to paper. Yet in Canada we have neither a comprehensive history of iron and steel production nor a complete history of steelworkers.[4] This book is a step toward filling that gap. It covers the period between the 1880s and the 1930s and is divided into four parts. The first will outline the main structural dynamics of the Canadian steel industry, which determined the terrain on which steel companies and their employees confronted each other, that is, business organi- zation and market pressures that prompted particular managerial and technological decisions. The second will present the technologi- cal transformation of the Canadian industry after 1900 and its implications for the work force. The third will assess the labour policies developed by Canadian steel companies to apply as effec- tively as possible the labour power they purchased from their work- ers. And the fourth will consider these workers' responses to the new work environment of the steel mills and their efforts to assert their own needs and concerns.

The available primary sources limit somewhat the scope of such a study. Canada's steel companies have not kept many records from their early days (the Steel Company of Canada will not even let anyone see what it has). Fortunately, it is possible to piece together information from a variety of other sources. Federal and provincial governments gathered some useful statistical data in the early twentieth century, and several commissions and boards of inquiry touched on working conditions in the steel industry, though there were never any intensive studies to compare with those undertaken by social investigators and state agencies in the United States.[5] One of the best windows on the world of work in this period is the business or trade journal, which often carried detailed articles about the new organization and technology in Canadian steel plants. This business journalism, however, must be approached with more caution than it often has been in the past. It was not straightforward, objective reporting, but rather a kind of promotional literature used by corporations to emphasize their modernity, efficiency, and steady control over their production processes, to build confidence in their enterprises among other capitalists, and to consolidate support in the wider Canadian public. These were not occasions for admitting the limitations or problems of their technology, the continuing necessity for skill and experience, or any kind of reliance on their workers. These articles were, in short, ideologically tainted and must be used carefully, in conjunction with other, less public sources. Private correspondence and testimony before government commissions can provide revealing qualifications to the technical articles; so, too, can oral history with former steelworkers who worked in the same plants, few of whom are still alive, but some of whom were interviewed during the 1970s. Finally, for the most difficult part of the story to unravel, the aspirations and activities of the steelworkers themselves, the daily newspapers in each town and the local publications of the labour movement can be rich sources.

A word is in order about what the book will not do. In the absence of more detailed sources, it can provide only the broad strokes on the canvas. In particular, it is not possible to undertake a careful analysis of individual work groups on the shop floor, which so many industrial sociologists have quite correctly and fruitfully highlighted in their studies of workers on the job. Any full understanding of workplace transformations in twentieth-century Canada must recognize the persistence of this nuclear form of solidarity and self-activity. Secondly, the focus on the workplace in this book necessarily downplays the community life of steelworkers. It can only begin to hint at the powerful ideological, cultural, and material forces within working-class communities outside the plants – the solidarities of kin and neighbourhood, the ethnic identities, the religious associations, the informal social networks like street gangs, or the influence of radical politics. A complete assessment of the remaking of the Canadian working class in the period between the turn of the century and World War II would have to take account of this life outside the workplace in more detail than was possible here. The assumption in this book, however, is that Canadian steelworkers' social and cultural life evolved in the larger context set by the world of work, which could severely limit the range of possibilities for individual workers, their families, and their communities.

In essence, this book will focus on what the emergence of mass-production, of the so-called "Second" Industrial Revolution, meant for the Canadian working class. We will be looking over our shoulders to see what was left behind in the nineteenth century, but also far ahead to see what the developments of this period would mean for twentieth-century working-class life. Today Canadians are deeply concerned about the implications of the rapid, disruptive technological change of the 1970s and 1980s. A study of the rise of one of Canada's most important mass-production industries can give us a valuable perspective on what is lost and what is gained when workers confront dramatic transformations in the workplace. Perhaps it can provoke some new questions about that kind of process. A social historian should hope for at least that much relevance in a serious dialogue with the past.

I

Corporations

Iron is one of the most common minerals on the earth's surface. For centuries men have found ways of transforming it from its natural state into usable tools. Before the last third of the nineteenth century, the products of primary ironworking were either brittle *cast* iron, which was poured into moulds to make metal objects like pots and kettles, or more malleable *wrought* iron, which had to be reworked with a hammer or run through a set of rolls into bars for use as nails, ploughs, or whatever. Cast iron was used by the moulder, wrought iron by the blacksmith. Small quantities of a stronger iron product with higher carbon content, known as *steel*, had been produced, especially for knives and other cutting tools, but only in the second half of the nineteenth century did the particularly strong metal we now know as steel come into large-scale production in Europe and North America.

Most countries with aspirations to industrial greatness have always wanted an iron and steel industry. Since the beginning of the first Industrial Revolution in eighteenth-century Britain, the production of iron and its various by-products has held out the promise of diversified manufacturing. Iron and steel became crucial materials for machinery, steamships, railway equipment, automobiles and other land vehicles, bridge and building structures, and countless consumer products. So the country able to assemble iron and steel facilities could hope to make it into the big league of industrial nations.

The development of this industry, however, followed different paths in different parts of the world. The particular combination of natural resources, capital, labour, and markets resulted in distinctive features in the scale, technology, and variety of production and in the structures of control.[1] Early Canadian iron production fell into three phases. Before 1850 small independent iron works opened on the site of the ore deposits and produced primarily cast iron to be turned into consumer goods for local settlements. Wrought iron used in British North America was mostly hammered out in the small forges of artisanal blacksmiths in hundreds of colonial towns and villages. In the second half of the nineteenth century, iron production was still dispersed throughout the countryside, but was re-oriented to the new markets created by industrialization, especially in the new railway industry. The "finishing" end of the industry, that is, the new rolling mills for shaping wrought iron, remained independent of the primary iron works during this period. Only at the turn of the century did the industry undergo consolidation, integration, and expansion of productive capacity. And only then did steel become the main product of the industry.

In 1896 the manager of Nova Scotia's Londonderry Iron Company raised few eyebrows when he lamented that "there is only one country under the sun with 15,000 miles of railway that does not manufacture a single steel rail, and that country is Canada."[2] Canadian businessmen could only nod soberly in agreement that the history of iron and steel production in the country up to that point had been largely a tale of failures and disappointments. This sorry record did not reflect any lack of initiative. The French government had authorized the construction of the first iron works in what was to become Canada as far back as 1730. After a brief interruption, that enterprise, the Forges de Saint-Maurice, continued in production for the next 150 years, during which time some two dozen similar projects came and went throughout central and eastern British North America.[3]

The most persistent and intractable problem was the quality and availability of the resources. Over and over again, iron men struggled with iron ore deposits that had annoyingly high sulphur and phosphorous content and that had often proved too limited in size. Blast furnaces were frequently abandoned after only a few years in the wake of fruitless metallurgical experiments. Charcoal was no problem until the forests had been denuded, but, as coke became the more common substitute in pig-iron production in the second half of the nineteenth century, the central Canadian iron manufacturers had to begin importing coal and those of the Maritimes had to try coking their troublesome local coal supplies. By the 1880s most of the stable ironmaking operations in Canada had settled down to producing charcoal iron, a specialized form of pig iron with increasingly limited uses (mostly for railway wheels) in a period when steel was emerging as the most popular new metal.[4]

A second, related dilemma was rooted in the labour process itself. As we will see in Chapter Two, nineteenth-century iron and steel production still required the skills of knowledgeable artisans who had mastered the "arts and mysteries" of metallurgy. Outside of the long-established community of Saint-Maurice, where skills were passed on from generation to generation, this know-how was scarce in Canada and had to be imported. The wages demanded for this artisanal competence were reputedly higher than elsewhere and pushed up production costs. Yet the difficulties created by the poor-quality ores made these men essential to the potential success of most iron works.[5]

The third major problem was markets. The Canadian market for iron and steel was relatively small and was inundated with cheaper British and American goods, manufactured by the 1860s and 1870s in the more technologically advanced production processes that were giving the world the first new steel. At this point, the structure of the Canadian iron industry, with separate primary and secondary producers, worked against the blast-furnace owners. A few of the iron works included small rolling mills for producing bar iron, but most of their pig iron found its way to independent foundrymen and rolling-mill owners in the larger industrial centres, especially Saint John, Montreal, Toronto, and Hamilton. These manufacturers wanted inexpensive pig iron and did not believe that the weak Canadian primary industry alone could meet their needs. Many of the rolling mills, in fact, rolled scrap iron rather than locally produced pig iron.[6] In 1879, when Sir John A. Macdonald's Conservative government sounded out business opinion on the specifics of the National Policy tariff structure, the only iron producers were three small furnaces in Quebec and a wobbly operation at Londonderry, Nova Scotia. The consumers of primary iron and steel, therefore, managed to keep the duty on these products low. The railway companies in particular made sure that steel rails stayed on the free list. These interests continued to fight for lower levels still, and in 1883 the federal government resorted to the alternative protective policy of paying bounties for iron produced in Canada. Only once the primary producers had proven their ability to meet Canadian needs on specific products would the tariff be raised, as in the case of steel rails in 1903. Right through the 1920s, the tariff structure for iron and steel products would remain riddled with gaping loopholes through which secondary producers could drive their trainloads of pig iron and secondary steel goods. The National Policy that gave life to so many other manufacturing industries was ultimately no great friend to the primary iron and steel industry in Canada.[7]

It is not surprising, then, that by the 1890s the Canadian iron and steel industry was a small, fragile edifice by comparison with the huge, highly centralized, and more integrated complexes in Europe and the United States, where transformation of production processes was resulting in much cheaper goods. Canadian iron works were scattered, antiquated, and generally isolated from the finishing sector of the industry, the rolling mills. Yet amid the gloom, that decade also saw several new entrepreneurial initiatives to overcome the obstacles to primary production and to implant a world-class steel industry on Canadian soil. Those efforts would follow two quite different paths.

The Big Four

By the turn of the twentieth century, four groups of capitalists were determined to exploit the apparently insatiable appetite of Canadian industry for steel and were working out different means of profitable survival in the peculiarly difficult economic and social climate. Their entrepreneurial options became either prudence or flamboyance, and their efforts created Nova Scotia Steel, the Steel Company of Canada, Dominion Iron and Steel, and Algoma Steel.

The Nova Scotians pioneered the strategy of slow, careful expansion based on a thorough familiarity with ironworking and on diversified markets in many different metal-working and transportation industries. In 1872 two New Glasgow blacksmiths, Graham Fraser and George Forest McKay, opened a small shop first known as the Hope Iron Works, and soon renamed the Nova Scotia Forge Company, with ten men to help them produce forgings for ships and railways. Six years later, the small firm moved two miles outside the town to the new industrial suburb of Trenton. In the National Policy hothouse of the 1880s, the demand for the firm's products, and the widening markets for iron and steel generally, encouraged the same New Glasgow men to organize a separate company to manufacture open-hearth steel on a modest scale. The new Nova Scotia Steel Company turned out the country's first steel ingots in July, 1883. In 1889 the two firms amalgamated to form the Nova Scotia Steel and Forge Company, still a locally controlled enterprise run in large part by the Fraser family. The next year the shareholders of this prosperous firm began a process of vertical integration by forming the New Glasgow Iron, Coal and Railway

Company, which built a short line to connect with the Intercolonial and with the company's newly acquired ore deposits in Pictou County. In 1891 it erected a large new blast furnace and a battery of the most modern coke ovens a short distance away at the new industrial centre of Ferrona. This new works survived the economic doldrums of the 1890s reasonably well, shipping pig iron as far away as Ontario, and another merger brought the two New Glasgow firms together as the Nova Scotia Steel Company in 1895.[8]

Thus far, "Scotia" had been a relatively small, integrated firm centred in Pictou County, Nova Scotia. But the inescapable problems with local ore and coal supplies sent the owners further afield to explore more reliable resources. They experimented with the recently discovered Wabana ore deposits on Bell Island off Newfoundland and in 1894 bought them up. In 1900 they also purchased the Cape Breton coalfields of the old General Mining Association and a year later reorganized their firm as the Nova Scotia Steel and Coal Company. Transporting these more distant resources inland to Ferrona soon proved too awkward and inefficient, and the directors decided to lay out large new facilities to produce pig iron and steel billets on the north side of Sydney Harbour, close to the coal and within easy reach by water of the iron ore. In 1904 the Ferrona blast furnace and Trenton open-hearth furnaces were therefore dismantled. The steel produced at Sydney Mines, however, still travelled back to the Trenton finishing mills, where, according to a Toronto *Globe* reporter, "over 700 varieties" of metal products were being turned out by World War I. In 1912 Scotia's directors launched an adjunct firm, the Eastern Car Company, to absorb more of the Trenton output and to attempt to cash in on the pre-war railway boom. During the war the corporate net was widened to include shipbuilding yards at New Glasgow. By this point, moreover, the corporation had long since lost its Pictou County flavour, first through the infusion of capital from Halifax's regional magnates, and then through a takeover by American capitalists in 1917.[9]

In the four decades since its modest beginnings in an artisanal workshop, Scotia had not only grown to major proportions, amassed resource holdings and manufacturing facilities worth millions, and assembled a steelmaking work force of over 2,000; it had also won a solid reputation for its technological innovations – the first open-hearth furnaces, "coal-washing" facilities, retort coke ovens, hydraulic forge, and so on. Not surprisingly, then, in 1914 the federal government's Shell Committee assigned the firm the task of developing the specialized steel needed for munitions production, which had previously not been made in Canada.[10]

By this point, Scotia had won its place among the "big four" Canadian steel producers by the careful, efficient development of an enterprise with well-diversified markets. "To a very considerable degree the success that has attended the company is due to this varied nature of its operations," a Canadian journalist explained in 1912. Shortly afterward, the first serious student of the Canadian iron and steel industry, W.J.A. Donald, reached the same conclusion: "Since all lines of production are not apt to be affected by competition and low prices at the same time, profits are less liable to extreme fluctuation," he wrote.[11]

Scotia was not alone in pursuing this strategy of diversification. On December 30, 1895, an exuberant crowd of several hundred visitors cheered the "blowing in" of the new furnace of the Hamilton Blast Furnace Company, a project backed by the leading capitalists of "the Birmingham of Canada" to produce pig iron and, as soon as possible, steel. The iron began to flow within a few weeks, but steel production had to wait for the company's amalgamation in 1899 with a twenty-year-old Hamilton enterprise, the Ontario Rolling Mills Company. The next year the new corporation, now known as the Hamilton Steel and Iron Company, opened a small steel plant. Over the next decade expansion was cautious, but the corporation did an increasingly lucrative business in foundry iron and secondary iron and steel products for both the railways and the general southern Ontario market. Two more open-hearth furnaces and equipment to produce railway spikes were gradually added, and in 1907 a second blast furnace, but the corporation's great leap forward came in 1910.

That year the Hamilton Steel and Iron Company began to discuss a merger with two Ontario corporations in the finishing end of the iron and steel business, the Canada Screw Company in Hamilton and the Canada Bolt and Nut Company with plants in Gananoque, Belleville, Toronto, Swansea, and Brantford. This amalgamation would have brought together most of the province's major secondary iron and steel producers. The merger plans were expanded, however, when the Ontario men were contacted by the energetic young promoter, Max Aitken, president of Royal Securities Corporation and the architect of the most ambitious merger in living memory, Canada Cement. Aitken had just purchased the large Montreal Rolling Mills, which he had tried unsuccessfully to sell to Dominion Iron and Steel in Nova Scotia, and which he had refused for patriotic reasons to sell to US Steel. He now proceeded to arrange an amalgamation of the three Ontario firms, his own Montreal company, and the large Montreal-based Dominion Wire Manufacturing Company, which its owner, US Steel, had decided to unload. The new corporation that emerged was the Steel Company of Canada (Stelco), controlled and managed by the Hamilton interests but including prominent Toronto and Montreal capitalists on its board of directors. It had brought together extensive facilities to produce a wide range of metal products – screws, nuts and bolts, wires, nails, pipes, and much more – but only limited capacity for supplying these plants with primary steel. A massive expansion of the corporation's production facilities in Hamilton therefore followed the merger: two new open-hearth furnaces and large new rolling mills capable of producing blooms, billets, rods, and bars, which opened in 1913. The demand for munitions in World War I brought a further increase in capacity and the addition of by-product coke ovens and a mill for the production of sheet metal, as well as vertical integration into iron-ore and coal-mining operations in the United States. Stelco's rise had been slow, and, in contrast to the other major producers, had proceeded through corporate mergers. But, by the end of World War I, it was undoubtedly the industry's greatest success story.[12]

During these same years, there was another radically different choice for would-be steelmasters. Rather than attempting to satisfy many different industrial requirements, a firm could adopt the American model of turning out large quantities of the same, highly specialized product. High volume meant fewer costly changeovers and permitted economies of scale. In the midst of the early twentieth-century railway boom, one product above all others looked promising – steel rails. And two groups of capitalists accordingly proceeded to lay down huge specialized steel plants geared to mass-producing the country's "ribbons of steel." In contrast to the New Glasgow and Hamilton men, however, these entrepreneurs were freewheeling financiers with investments in natural resources, rather than experienced iron men. That difference would prove crucial to their relative success.

The first of these dramatic new projects arose in the sleepy little town of Sydney in Cape Breton and grew out of "the genius, the magnetic and irresistible enterprise of one man," Henry Melville Whitney, a New England businessman who had organized the Dominion Coal Company in 1893 to consolidate the ownership of a huge proportion of the Nova Scotia coalfields. In 1899, after

Nova Scotia Steel rebuffed Whitney's overtures for a merger, he launched the Dominion Iron and Steel Company (Disco), with a board of directors comprised of his Dominion Coal colleagues and a handful of Toronto and Montreal's leading capitalists. With easy access to Bell Island iron ore, Cape Breton coal, and ocean transportation routes, Disco's promoters envisaged supplying the Canadian market as well as exporting to Europe in successful competition with American, British, and European producers. According to the *Canadian Mining Review*, the Sydney steel works would "exceed in magnitude and capacity any individual concern on this continent." A British business journalist believed that no Canadian enterprise had been started "with a greater flourish of trumpets or a fairer promise of success on a large scale," and noted that steelmakers on the other side of the Atlantic were quaking in their boots: Disco "seemed like holding a suspended sword over the iron trade of the Mother Country."[13]

The optimism surrounding this venture was boundless. "I cannot control my enthusiasm when I think of the future," Whitney told a Sydney audience. And a writer in the *Montreal Daily Star* saw vast benefits for "the whole Canadian people": "The building up of a big manufacturing city at the front door of Canada will give our Dominion new importance in the eyes of the world. There are very few countries that export iron and steel, and if Canada can take rank among the iron exporting nations it will add greatly to Canadian prestige." As the unprecedentedly large, modern, $15-million complex reached completion in 1902, the Canadian Manufacturers' Association publication, *Industrial Canada*, described it as "the outstanding feature of our industrial development of the past few years." To the casual observer, it looked like Canada had finally made it into the big league of world steel producers.[14]

Before long, however, the cracks were appearing in this magnificent edifice. The grand visions too often translated into extravagant expenditures. Early estimates of production costs proved ridiculously low. Cost accounting became chaotic. Managerial staff came and went too frequently. More importantly, Whitney and his associates made expensive blunders that reflected their ignorance of steel production and increased the financial burden: Bessemer converters had to be abandoned when the local ore proved inappropriate for the steelmaking process; and a half-built rail mill was dismantled and replaced. The *Canadian Mining Review* even accused the Canadian financiers in control of the firm of being more interested in stock-market manipulation than in efficient steel production. In 1903 the same journal summed up the corporation's first four years as a history of "kaleidoscopic changes, of extravagance, vacillation and blundering."[15] By the end of that year, staff was being laid off and wages slashed, and for several weeks the next summer the firm had to weather a bitter strike by its angry workers. By the end of 1904, the ambitions of the Disco directors had narrowed down to serving only the domestic market. At first, the only finished steel products the corporation was producing came from the wire mill, which opened in 1904. The rail mill finally started up in June, 1905, and production increased steadily in the pre-war years. A prolonged battle with the Dominion Coal Company ended with the creation of a new holding company in 1910, the Dominion Steel Corporation. In 1911-12 Disco expanded its facilities and made a modest effort at product diversification. After a rocky start, therefore, Disco had established itself by the outbreak of the war as the country's leading steelmaking enterprise, if not quite living up to the grandiose turn-of-the-century dreams. Like the other major steel companies, it enjoyed a new boom during World War I.[16]

Meanwhile a similar drama was unfolding a thousand miles inland. In 1893 another American entrepreneur, Francis Henry Clergue, arrived in the quiet northern Ontario town of Sault Ste. Marie to investigate a hydroelectric power site on behalf of some American investors. Clergue already had a remarkable track record as an ambitious promoter of ill-fated power and transportation projects in Maine, Alabama, and Persia, but the Sault project fired his imagination anew. By the early 1900s he had expanded the original plans to include a massive pulp mill, an associated sulphite pulp plant, iron and nickel mines, a nickel refinery, street railways on both sides of the international border, vast tracts of timber leases, the Algoma Central Railway, and huge chunks of valuable real estate – all of which were assembled under the umbrella of the Consolidated Lake Superior Corporation. It was the discovery of iron ore in the Michipicoten region and the opening of the rich Helen Mine in 1899 that spurred the irrepressible Clergue on to the still grander vision of a major new iron and steel complex. With ambitious plans to seize the new market for steel rails in the Canadian West, he and his American associates incorporated the Algoma Steel Company within the larger corporation in 1901. Clergue's infectious optimism soon spread to the civic leaders of "New Ontario," the Toronto business community, and the Ontario Liberal government. His daring and imagination similarly impressed a young engineer in the firm, Alan Sullivan, who would later try to mythologize his employer in his novel *The Rapids*.[17]

Clergue's dream, however, quickly suffered the same fate as Whitney's, for many of the same reasons. His decisions in building a steelmaking complex were certainly questionable. In the first place, Sault Ste. Marie was far from an ideal location. Not only was it too distant from most markets for iron and steel, but it contradicted the new wisdom among North America steelmakers to build plants closer to the fuel supply and when necessary to ship in the ore, which is lighter and cheaper to transport. Clergue decided to rely on his cheap hydro and local charcoal resources, at a time when the use of charcoal was rapidly declining. A costly conversion to the use of coke would eventually be necessary, along with the inefficient importation of that product until coke ovens were constructed in 1910. A misjudgement on a similar scale was the decision to construct steel furnaces and a rail mill before erecting any blast furnaces. The corporation therefore had to bear the added cost of shipping in pig iron, mostly from the iron works at Midland, Ontario (of which Clergue was a director). An equally serious blunder was the installation of Bessemer steelmaking facilities, rather than the increasingly more popular open-hearth process. The iron ore in Clergue's celebrated Michipicoten mines proved extremely difficult to use in the Bessemer converters, and soon American ore had to be imported as well. Eventually, in 1907, open-hearth furnaces were built (for which the appropriate pig iron had to be imported at first) and in 1916 the last Bessemer converter was shut down. Seldom has a major industrial enterprise been planned with as much faulty logic and glaring incompetence. Small wonder, then, that the plant had to be closed in December, 1902, only six months after the first rail was rolled, since it could not compete with German producers for a Canadian rail contract.[18]

As if the technical blunders were not enough, Algoma Steel was caught in the corporate quagmire of Consolidated Lake Superior. Clergue's enthusiasm for expansion had required constant infusions of new capital and a massive debt soon accumulated. An internal shake-up that forced the energetic president out of office in April, 1903, could not head off the crash that hit in September. After months of negotiation over reorganization and refinancing, operations ground to a halt when the firm found it could not meet its payroll. The financial disaster became a social crisis ten days later, when hundreds of angry unemployed workers converged on Sault

Ste. Marie to demand the wages owed to them. Crowds stoned the corporation's offices and its office staff and dispersed only when the militia was called in from Toronto. Ontario's Liberal government, which had identified itself so closely with this flagship of northern resource development, rushed to the corporation's aid with funds to meet the payroll and a $2 million loan, and along with the major creditors insisted on an internal reorganization.[19]

Out of this morass emerged a new holding company, the Lake Superior Corporation, which was in more cautious hands but which inherited the financial burden of Clergue's spendthrift days. Ownership of the corporation lay in two distinct blocks of shareholders, one in Philadelphia and the other in England, between whom control shifted periodically. Meeting the firm's obligations to these absentee owners would hamper capital expenditures on expansion until the 1930s.[20] In the decade before the war, various appendages to the corporation were sold off, and Algoma Steel became the centrepiece of corporate planning. The steel plant started up again in the summer of 1904, and, with the addition of blast furnaces in 1904-1905 and further expansion, renovation, and vertical integration into mining between 1910 and 1914, the corporation increased its production enormously. Despite some meagre efforts at diversifying its output just before the war (to permit production of railway track fastenings), the main product remained steel rails. Like the other three major Canadian steelmakers, however, Algoma threw itself into highly profitable munitions work during the war and was able to increase its productive capacity and to begin planning a new mill to turn out structural steel. And like its corporate counterpart in Cape Breton, Sault Ste. Marie's steel complex seemed to have recovered from its disastrous infancy.[21]

Scotia, Stelco, Disco, and Algoma were not the only new ventures into iron and steel production in the early 1900s, but virtually all the others had collapsed by World War I. Several groups of investors sank their money into plants in Ontario at various points along the Great Lakes waterfront – Deseronto, Midland, Collingwood, Parry Sound, and Port Arthur – but none of these could cope with the difficult market situation in Canada.[22] Probably the most ambitious project began in Montreal, where the Drummond family ran a large foundry business, and included blast furnaces at Midland, Ontario, and Londonderry, Nova Scotia. In 1908 the Drummonds created a new Canada Iron Corporation to pull together their motley collection of investments in foundries, machine shops, blast furnaces, and iron mines, which Nova Scotia Steel's general manager accurately described as an "aggregation of more or less worn out, decrepit, and unpaying properties." Five years later the corporation passed into receivership, and its iron works in Nova Scotia, Quebec, and Ontario were never reopened.[23] The only survivors alongside the "big four" were the small, American-owned Canadian Furnace Company at Port Colborne, a few small specialized firms in Ontario and Quebec using the expensive electric steelmaking process, a handful of independent rolling mills, like the Manitoba Rolling Mills in Winnipeg and Burlington Steel in Hamilton, and a small number of independent steel foundries, notably Dominion Foundries and Steel (Dofasco) in Hamilton and Canadian Steel Foundries at Montreal and Welland, which manufactured small quantities of steel. Aside from the Canadian Furnace Company, none of these were primary producers with blast furnaces.[24]

By the end of World War I, then, the Canadian steel industry had expanded phenomenally from the gloomy days of the 1890s. The national output of pig iron had risen from 38,000 long tons in 1895 to just over over a million in 1918, and steel ingots and castings had shot up from 17,000 to 1.67 million over the same period.[25] More importantly, the fragmentation and dispersion of the nineteenth-century industry were gone. The bulk of the country's 20,000 steelworkers held jobs in the four major, integrated corporations that now turned out most of the primary and secondary steel produced in the country. The war had pulled the industry out of the serious slump of 1913-15 and boosted it to unprecedented heights of performance. With the signing of the Armistice, however, a dark cloud hung over post-war prospects.

A Fragile Structure

By September, 1919, the Canadian steel industry was in a serious slump. All the blast furnaces in Nova Scotia had been shut down, and the Ontario plants were operating at no more than half capacity. In surveying the situation, the industry's new trade publication, *Iron and Steel of Canada*, reminded the major producers of the hazards of returning to pre-war production patterns: "One fault of the larger steel companies of Canada has been their reliance on one main product, necessitating large tonnage production for profitable operation, and somewhat out of balance with the requirements of Canada itself." The journal cautioned against aiming at an export market and urged diversification to permit "the manufacture of steel in such grade and of such size and shape as will foster the evolution of numberless small trades that should grow out of a basic iron and steel industry." This seemed to be the collective wisdom in the steel company boardrooms as well, but the big four were not all equally capable of adapting and surviving profitably.[26]

Algoma was probably the most vulnerable of the big four. It had turned out the largest tonnage of shell steel, but had not overcome the narrow specialization in rail production that had characterized its development before the war. An interest-free federal loan had helped the firm complete the shift from Bessemer to open-hearth steelmaking, and early in 1920 construction began on a new structural steel mill on the expectation of assured government orders. The state's largesse never materialized, however, and construction of the mill was abruptly halted a few months later. The corporation's ability to adapt on its own remained severely constrained by the financial quagmire of the Lake Superior Corporation's continuing indebtedness. The absentee bondholders and stockholders still disagreed about development plans, and the American interests that had control by the end of the war insisted on milking profits from the existing production facilities rather than diversifying on a major scale. This proved to be an ill-conceived strategy since the days of large railway orders were past. During most of the 1920s and 1930s Algoma operated at well less than half its rated capacity, often shutting down for months at a time between sporadic orders from Canada's overbuilt transportation lines.[27]

The same prospect was facing Disco's similarly overspecialized operations in 1919, but the attempted solution was much more spectacular. That year control of the company passed to two dominant blocks of investors, one Canadian and one British. At the same time, the American interests that had bought control of Nova Scotia Steel in 1917 approached Disco's new owners with proposals for a merger. These discussions eventually reached fruition in 1920 with the creation of the British Empire Steel Corporation (Besco), a massive consolidation of both Nova Scotian steel companies, extensive coal holdings, and the Halifax Shipyards. But the heavily watered stock placed crippling fixed charges on the new firm, and accusations of gross mismanagement by officials unfamiliar with the industry arose from many quarters, just as they had twenty years earlier. The economic depression of the early 1920s further constrained the new corporation's profit-making potential, as did the much higher freight rates introduced by the new Canadian National Railways in 1920 to end this longstanding concession in transportation policy to Maritime industry. A ship-plate mill, opened with great optimism in February, 1920, had to be closed when the federal government cancelled the orders it had promised. In 1926 Besco went into receivership and two years later was reorganized by central Canadian capitalists. For most of the 1920s and 1930s the Sydney steel plant still operated on the same restricted, hand-to-mouth basis as its counterpart in Sault Ste. Marie.[28]

Disco also managed to drag down the once widely respected Nova Scotia Steel Company, but the roots of Scotia's difficulties ran deeper than the immediate crisis of corporate mismanagement. In the decade before the war, Stelco had begun to threaten the company's access to the central Canadian market, and the post-war changes in freight rates made competition with Ontario steelmakers more difficult. The Maritime metal-working industries were not large enough to take up the slack and immediately after the war began to show signs of long-term decline and decay, as central Canadian industries closed down their Maritime operations or drove local competitors out of business. Partly to compensate for the loss of these markets, Scotia's pre-war expansion plans had involved a shift toward greater production for the transportation industries. The opening of its subsidiary, the Eastern Car Company, was clearly part of this reorientation. In 1919 the corporation's general manager at Trenton told a royal commission: "Our work for a number of years has been largely railway work. . . . Our orders vary very materially according to the needs of the railroads." As Algoma and Disco were discovering, however, this was an unreliable basis for industrial growth. In November, 1920, Scotia's blast furnaces and open-hearth facilities at Sydney Mines were shut down, never to be relighted. As part of Besco's rationalization schemes, Trenton's "cogging" (blooming) mill was closed in favour of Sydney's plant. All that remained of Scotia's carefully developed iron and steel complex were Trenton's increasingly antiquated finishing mills and car works, which, like the Sault Ste. Marie and Sydney operations, limped on through the 1920s and 1930s with few orders and widespread unemployment. The "big four" had shrunk to three.[29]

The fundamental problem for the Nova Scotian and Sault Ste. Marie corporations was that the demand for steel goods was shifting so dramatically from railways to automobiles and mass-production consumer goods. The absence of rolling mills suited to this new demand was enough of a problem for Algoma and Besco, but, as the regional specialization of the Canadian economy increased, the newer industries were almost all in southern Ontario, far from Sault Ste. Marie, Sydney, or Trenton. Excessive transportation costs made steel from these mills even more expensive than products from Buffalo, New York.[30] Not surprisingly, then, the only success story in the post-war Canadian steel industry was at Stelco. Like its competitors, the Hamilton-based corporation had to cope with the excess capacity built up during the war and the limited demands of a depressed economy in the early 1920s. But the firm was in a much better position to handle these pressures. Its production facilities were much more diversified and, perhaps most important, were much closer to the heaviest consumers of primary and semi-finished steel in the southern Ontario industrial heartland. Prudent management decisions had begun the transition to post-war production well before the end of hostilities, and the corporation applied the considerable profits accumulated in munitions work to a substantial upgrading of the technology in its many plants. By the mid-1920s Stelco had emerged as the new giant of the industry. It remained in such a healthy financial position in the difficult interwar period that profits continued to mount up and dividend payments were never once suspended.[31]

Given the evident collapse of so much of the industry, all the steel companies might legitimately have expected some sympathetic attention from the Canadian state. For nearly half a century there had certainly been no lack of interest at all levels of government in promoting a domestic iron and steel industry. Municipal councils had showered the new corporations with tax concessions, land grants, and bonuses. Federal and provincial governments had initiated bounty payments to primary producers in the 1880s and 1890s, although these stimulative programs had all been wound down by 1912.[32] Yet the federal tariff structure for steel was more problematic. It contained huge loopholes and provisions for payments of drawbacks to many consumers of the industry's primary products who imported their iron and steel. After Francis Clergue had convinced the Laurier government in 1900 of the economic necessity and political expediency of a higher tariff on steel rails, Algoma and Disco had found a market in Canada for this highly specialized production. But in most categories of steelmaking, especially the increasingly important structural steel and sheet-metal work, cheap imports dominated the Canadian market. By the latter half of the 1920s, 57 per cent of the primary steel consumed in Canada was imported. Steel producers generally did not benefit from the Canadian state's commitment to industrialization through import-substitution that gave life to so many other manufacturing industries.[33]

Pressure on the federal government for upward revision of the steel tariff increased in 1926 when the King government appointed the Advisory Board on Tariff and Taxation. Both Algoma and Besco repeatedly presented plaintive appeals for help, backed by extensive documentation and promises to diversify their production. They were hampered by the relative indifference of Stelco, which was succeeding well without tariff adjustments, and by vocal opposition from a host of iron and steel consumers. Eventually, in 1930, the Liberals introduced increased protection for the industry in order to win back the political support they had lost in Nova Scotia, and later the same year the new Conservative government pushed the tariff rates still higher. Although imports declined, however, the economy was passing through its worst crisis in decades and the structural problems of two of the three steel companies remained to haunt their owners and managers. Disco and Algoma went through major corporate reorganizations in the 1930s, but it would take the return of another war and the active intervention of the federal government to transform these firms into hardy competitors in a continental market.[34]

The Canadian steel industry that emerged before World War II was thus a strange beast. It was controlled by a small handful of powerful corporations and therefore fit both the pattern of corporate consolidations in Canadian industry in the period and the organizational structure of the American industry. Yet the unique context of inadequate raw materials, a small domestic market, intense foreign competition, and weak tariff protection made the development of steel production a much more hazardous proposition than that facing many other Canadian industrialists in these years. The difficulties were compounded by the serious flaws in some of the entrepreneurial decisions that eventually brought the industry to life. Inexperience, incompetence, and financial manipulation burdened two of the major producers with serious long-term financial problems and resulted in large, overspecialized, underused facilities languishing in isolated locations far from the most important markets. An air of crisis and insecurity thus hung over large parts of the industry in the first four decades of the twentieth century. It remained a set of incomplete fragments that never succeeded in supplying more than a limited portion of the Canadian demand for iron and steel products.

Steelworkers and Their Industry

The oddly truncated, fragmented shape of the Canadian steel industry had serious implications for the men who worked in the plants and for the labour process in which they were involved. The intense competition from American steel producers in the small Canadian market produced significant distortions. On the one hand, there was pressure to match American standards of productive efficiency in order to survive. In fact, since most of the directors of Canadian steelmaking corporations were financiers with little or no expertise in running a steel plant, they recruited the vast majority of their managerial and supervisory staff from the United States. American "mass-production" methods were thus carried directly into Canadian factories.[35] On the other hand, certain lines of production never appeared, or remained underdeveloped, and many rolling-mill facilities retained outdated technology because of the corporations' inability to benefit from economies of scale. "It is unfortunate," Stelco's president wrote to the corporation's shareholders in 1927, "that the slow progress of the steel industry in Canada does not offer, in some lines, greater encouragement in support of the installation of the most economical units of large capacity, such as are used in broader markets."[36] For the many steelworkers on piecework, the frequent changeovers required by small orders meant lost wages during set-up time and lower bonus payments.[37] Even more important, the special problems of Canadian steel production were also used to stall improvements in wages and working conditions, especially the abolition of the twelve-hour working day.[38]

Probably the most devastating consequence of the steel industry's fragility and instability was the chronic insecurity of employment and income for steelworkers. Steel plants normally operate on a non-stop, year-round basis, but, above and beyond the regular downturns in the business cycles that hit all Canadian industry every few years, two of the major steel producers had a much more erratic pattern of production throughout this period. The spasmodic start at Algoma and Disco in the early 1900s was followed first by heady booms before and during the war, and then by catastrophic curtailments of production for most of the 1920s and

1930s. In 1927 the federal Department of Labour reported that employment in the industry had risen to only 68 per cent of the 1920 level. The next year Algoma admitted to the Advisory Board on Tariff and Taxation that "the great majority" of its workers had worked only five to nine months a year since 1920. "The result of so many idle days in our Mills and consequent loss of time by employees, year after year, has resulted in our City experiencing a long period of general financial depression," a group of the firm's employees told the board. And, as these workers noted, this insecurity of employment produced transiency among the men, especially "a considerable exodus to the United States." For those who hung on in the steel towns, the large pool of labour could severely depress wages and increase competition for jobs. In the 1920s there were always dozens, sometimes hundreds, of anxious workers outside the plant gates at Algoma and Disco hoping to be hired. This kind of reserve army of labour was still present in Sydney in 1929 in the midst of a brief revival of business for Disco; the Sydney *Post* reported that steady work was available for only 2,800 of the 3,600 men on the payroll. "You couldn't spend money you earned," a retired Disco worker recalled, "because if you were going to get laid off, there was no unemployment insurance." In such circumstances steelworkers would live in fearful dependence on company officials for access to the precious jobs that existed.[39]

The geographical dispersion of the industry also created problems. The great distances separating the three main centres of primary production militated against easy movement between employers and against the development of any sense of occupational community among the industry's workers. There was no closely knit steelmaking area to compare with, say, western Pennsylvania in the United States.

The major steel centres were not even similar urban environments. New Glasgow-Trenton was a relatively small company town with deep roots in the past. It had grown from a flourishing shipbuilding centre in the nineteenth century to become the industrial hub of Pictou County, with a combined population of 7,800 by 1911. A few local metal-working firms continued to prosper in the pre-World War I boom, but Nova Scotia Steel clearly dominated the town's economic life. Despite the infusion of outside capital and its expansion to Cape Breton, Scotia had a reputation as a community institution that had built up a stable labour force of local residents and would protect their jobs. The fact that only the finishing processes of steel production were carried on in the town after 1904 also marked it off from the other major steelmaking centres, where large, ethnically mixed work forces were assembled to run the various furnaces.[40]

In contrast, Sydney's growth pattern was a much more abrupt explosion at the turn of the century, as the quiet port was suddenly transformed by a construction boom and a flood of new workers for the huge new steel plant, which pushed the town's population up from a mere 2,427 in 1891 to nearly 18,000 in 1911. The frequent comings and goings of managers and workers from the United States and Europe would give life in the town more exotic touches but also somewhat less cohesion than in most of Cape Breton Island or in New Glasgow-Trenton. The distant and impersonal corporation that soon came to dominate the town had to face a

Table 1

Population of Canadian Steelmaking Centres, 1891-1941

	Hamilton	Sault Ste. Marie	New Glasgow-Trenton	Sydney	Sydney Mines
1891	48,959	2,414	4,417	2,427	2,442
1901	52,634	7,169	5,721	9,909	3,191
1911	81,969	10,984	8,132	17,728	7,470
1921	114,151	21,092	11,818	22,545	8,327
1931	155,547	23,082	11,471	23,089	7,761
1941	166,337	25,794	11,909	28,305	8,198

SOURCE: *Census of Canada*, 1931, II, pp. 8-9, 12-13; 1941, II, pp. 9-10.

rich Cape Breton culture in and around Sydney that blended the well-established Scottish traditions of the rural population with the independent ways of the Island's coal miners. The proximity of large coal-mining communities, in fact, brought both the major Nova Scotia steel towns, along with Sydney Mines, into close contact with a set of distinctly different rhythms and social relations, which did not influence other Canadian steelmaking centres. The fact that by 1920 all the local steel plants and coal mines were in the same corporate hands further strengthened the regional consciousness and solidarity of Cape Breton's miners and steelworkers.[41]

Sault Ste. Marie parallelled Sydney's rapid growth and fluidity of population, but initially, at least, it had a more varied industrial base resulting from Clergue's grandiose vision. Aside from Algoma Steel, the major survivor after the collapse of Consolidated Lake

Superior was a sizable pulp and paper mill, but by World War I
steel was the town's chief industry. Perched on the edge of the
northern Ontario bush, the Sault remained much more isolated
than the other steelmaking centres. Algoma was therefore able to
exert a more effective control over the community and its work
force, even though it remained as much of an impersonal presence
as its counterpart in Sydney.[42]

Hamilton was worlds apart from the other steel towns. It was
vastly larger, reaching a population of 100,000 just before the war,
but, more importantly, it had a long history as a major manufacturing centre with an aggressive local elite intent on expanding the
stature of the city and its industries in the central Canadian economy. It was that elite that had launched the Hamilton Blast Furnace Company, and, even after the Stelco merger, control remained
in the city (until 1926 when the new president, Ross McMaster,
chose to work from Montreal). In the early twentieth century,
moreover, Hamilton was neither a company town like the other
three steelmaking centres, nor even a "steel" town, in which the
rhythms of urban life followed the fortunes of one central plant.
The so-called "Birmingham of Canada" had a much more diversified industrial structure, which included large clothing and textile
mills and several huge metal-working firms (notably Canadian Westinghouse and International Harvester), of which Stelco was not
even the largest. These industries brought together a similarly diverse work force, which included many skilled workers, along with
the same multicultural flavour that characterized Sydney and the
Sault. Stelco's other finishing plants were spread through similarly
mixed industrial communities, from Brantford to Montreal. Stelco's workers, then, were somewhat less pivotal to their cities and
towns, where they rubbed shoulders with a variety of other workers
in manufacturing whose concerns and cycles of industrial conflict
might not always coincide with those of the steelworkers.[43]

In general, then, Canadian steelworkers would find the structure
of their industry heavily determining their employers' managerial
decisions about technology, discipline, wages, and working conditions within the steel plants and seriously affecting both their living
standards and their ability to struggle toward some improvement.
From an international perspective, the work experience of Canadian steelworkers would resemble in many ways that of their
counterparts in the United States, but here the pressures on workers
in the industry would be even more intense.

2

Machines

It was a long walk in from the main road to reach the inside of a turn-of-the-century Canadian steel plant. The huge cluster of buildings usually stretched over acres of countryside on the far edge of the steel towns. Visitors' eyes always widened as they approached the plant and moved into the shadow of the belching, blazing smokestacks. Before they had stepped inside the plant, they would pause in awe at the massive proportions of the various furnaces and stoves and at the intricacy of the machinery that clattered around these fire-breathing monsters. No matter which door they chose to enter, their senses were assaulted by the foul, smoky air, the thunderous noise, and the general atmosphere of commotion, as the gigantic machinery lumbered along and small handfuls of men hurried to and fro in the dense gloom. But, above all, they would have left the plant with one indelible image from the steelmaking process – fire. In every department they met the fierce heat from furnace mouths, the blinding glare of molten iron and steel that often radiated showers of sparks, and the white-hot hunks of metal that the colossal machines devoured and spewed out. Many visitors later claimed they had stood at the gates of hell.

Craftsmen have been extracting iron from its ore and working it into usable products for centuries. The intense heat they applied always provided the drama that prompted so many observers to evoke the imagery of Dante's *Inferno*. The procedures used to apply this heat – and the nature of the end product – have nonetheless evolved considerably. In Canada, technological change in the iron and steel industry spanned three periods before World War II. In the pre-industrial era before 1850, ironmaking brought together fire, water power, and simple tools in the hands of brawny men. The move to production within an industrial-capitalist economy between the 1850s and the 1890s involved larger productive capac-

Handling iron ore at the Canadian Iron Furnace docks, Quebec.
(Canadian Mining Manual, 1897)

ity, steam power, some limited mechanization of the labour process, but a continuing reliance on human muscle-power. In the post-1900 era of highly mechanized, mass-production plants, huge new, electrically powered machinery came to dominate most branches of iron and steel production. Each of these phases corresponded with the development process in other industrial-capitalist societies, like Britain and the United States, but each began somewhat later in Canada. A closer examination of this evolution in iron and steel production in Canada will reveal how dramatically the labour process changed in the early twentieth century and will allow us to assess the impact of that transformation on skill requirements in the industry.

The Gigantic Automaton

The First Industrial Revolution

The few pre-industrial ironworks that operated in New France and British North America before 1850 were developed close to iron ore deposits and to rivers that could provide water power.[1] Iron-masters hired men to dig up the shallow deposits of iron and limestone and to fell hardwood trees for slow burning in special

"Beehive" charcoal kilns of the Canadian Iron Furnace Company, Grandes Piles, Quebec. (Canadian Mining Manual, 1897)

pits to produce charcoal. Often this work was contracted out to local farmers, as it was at the older charcoal-iron furnaces in Nova Scotia and Quebec until the turn of the century. The farmers' wagons would all converge on a work site dominated by the fiery stack of the blast furnace. Here the iron would be extracted from the ore to produce "pig iron." This conversion process has always involved the application of heat to a mixture of iron ore, charcoal or coke (the fuel for burning out the impurities), and limestone (a flux for congealing the impurities into a slag), with some draft (the "blast") to increase the heat. In these early years, the blast furnaces were tall, perhaps thirty-foot stone structures usually built beside a hill. A horizontal ramp from the hilltop allowed workmen to trundle the raw materials across in wheelbarrows and dump them into the top of the furnace. The "blast" of cold air blown into the bottom of the furnace would normally be powered by a simple waterwheel.

At regular intervals these primitive furnaces would be "tapped" by driving a bar into a clay stopper at the bottom of the furnace. A deafening roar and a burst of smoke and liquid fire would then erupt, producing a spectacular pyrotechnical display. The molten iron that then flowed out would either be ladled directly into castings (for stoves, pots, and so on) or be directed along channels in

In the casting house of the Deseronto Iron Company, around 1900, these workers would release molten iron from the base of the blast furnace and guide it along channels in the sand to form pig iron. (Ontario, Bureau of Mines, Report, 1908)

the sandy floor of the casting house to form three-foot blocks of iron known as "pigs." According to one observer, men scurried along the little rivers of fire "like the demons in a Kirafly spectacular theatrical hell, poking about with their flaming poles and seeing that each individual piglet was duly 'fed.' "[2] Then they shovelled sand over the molten iron until it cooled somewhat. Meanwhile, the furnace workers were hastening to plug up the tapping hole by ramming clay into it by hand. Since each pig was still connected to a central cord, labourers had to smash the blocks apart with hammers. The men then loaded them into wheelbarrows and carted them away. These methods could produce no more than five to ten tons of iron per day, and numerous labourers were required around the furnace with their barrows, shovels, and hammers.[3] The wrought iron was then produced by remelting and oxidizing the pig iron in special "Walloon hearths," or "fineries," and hammering the lumps of iron that emerged by hand or in a water-powered trip hammer.[4]

In all these stages of production, then, ironworkers applied their brawn and metallurgical knowledge, aided only by intense heat, water power, and some basic hand tools. After 1850, however, the ironmasters began restructuring their work processes to keep pace with the demands and pressures of the new age of industrial capitalism. Charcoal production was moved to special kilns by the blast furnace, and then, as the forests were depleted, gradually abandoned in the last quarter of the nineteenth century in favour of coke. Nova Scotia Steel introduced the most up-to-date coking facilities at its Ferrona iron works in 1892. The labour requirements were still substantial, however. Several men were kept busy shovelling coal into the ovens and shoving it out, hosing it down, and forking the still-warm product into wheelbarrows. Before 1900, moreover, it still took many men with shovels, wheelbarrows, and the occasional assistance of horses to move the raw materials up to the blast furnace. Even once these materials had to be imported from long distances, gangs of workers would be sent into railway cars or the holds of ships to shovel out the coal, iron ore, or limestone by hand.[5]

At the blast furnace itself, some important technical modifications had appeared by the 1890s, notably the introduction of specially designed stoves to heat the blast and steam engines to power it, which required new groups of workers to handle coal and to maintain the machinery. For most of the employees, however, work routines around the typical Canadian blast furnace had changed surprisingly little, even in the new plants at Ferrona and Hamilton. Steam power now enabled elevators to carry the wheelbarrows of raw material up the side of the furnace, but the brigades of shovellers and "barrow-men" still pushed their large, two-wheeled carts full of some 1,200 pounds of material from the piles by the railway tracks to weigh scales and then into the elevators. On a platform at the top, half a dozen "top-fillers" then grabbed the wheelbarrows and emptied the stock into the mouth of the furnace – a potentially dangerous job that cost one worker his life at the Hamilton Blast Furnace in the first year of its operation. At the base of the furnace, moreover, labourers still sweated in the heat and noise to direct the molten iron into the beds of pigs.[6]

The process of refining pig iron into wrought iron had gone through a much more significant transformation, however. In place of the blacksmith's hammering was a two-step process: puddling and rolling. The new procedure known as "puddling" had been developed in Britain in the late eighteenth century but had been slow to catch on on this side of the Atlantic. The reason for the delay was not hard to trace: iron puddlers were highly skilled men who were unknown in British North America until the small iron works and independent rolling mills of the 1850s and 1860s began importing small numbers of them. By the late 1890s, the Hamilton Blast Furnace Company was the only primary producer to run a few puddling furnaces, and these were closed down in 1907. Since most nineteenth-century Canadian rolling-mill operators preferred to use scrap iron and imported iron bars, puddling remained less significant than a similar process of working with scrap iron known as "bushelling."[7]

A puddler's trade demanded massive strength, tolerance of intense heat, and metallurgical know-how. He would throw pigs of iron into his small puddling furnace and stoke the fires while they melted. Then he would begin stirring ("puddling") the melting iron with a long rod inserted through the furnace door, adjusting the heat and adding iron oxide in order to drive out the excess carbon. At the moment when the charge in the furnace "came to nature," that is, when the pure iron and the slag were separating, he would use his pole to divide the pasty mass of iron into three equal balls of some 200 pounds each. These he would drag out of his furnace onto a buggy and wheel over to a device known as a "squeezer," through which he would run the iron balls to extract the last impurities. The man who ran a bushelling furnace did essentially the same work, only his raw material was scrap iron, not pig iron. All this heat and straining muscle-power had replaced the old hammering technique for changing the chemical composition of the pig iron to make it more flexible.[8]

For profit-conscious ironmasters, however, these craftsmen posed problems. They worked independently at their own pace and carefully guarded their trade secrets. Their output could not be increased without simply hiring more of them at the high wages that their skills commanded. Londonderry's nine puddling furnaces could turn out only 25 tons per day by 1890.[9] Not surprisingly, then, mill owners in North America and Western Europe eagerly turned to new production methods for refining iron that promised to by-pass the puddler and his little furnace and to turn out a versatile, cheaper iron product with a slightly higher carbon content – steel. The first of these new methods introduced in Canada was "open-hearth" production. In the large, brick-lined furnace, a gaseous fuel (usually a by-product of coke or pig-iron production) was blown in, along with air, to provide the heat for melting small quantities of scrap steel, limestone, and iron ore. Then molten pig iron would be poured in through the furnace door, and, during a five-to-ten-hour "heat," impurities in the iron would be driven off as vapour or drawn into a slag by the limestone. Once the furnace crew had shovelled in small quantities of ferro-alloys to adjust the chemical composition and had tested the brew, the slag and molten steel could be tapped from a spout at the back of the furnace in thunderous explosions of liquid fire. (Some of the smaller furnaces could be tilted to pour out the liquid.)

One of the British inventors of the process, C.W. Siemens, had tried unsuccessfully to make steel at the Londonderry ironworks in the 1870s. Nova Scotia Steel installed one fifteen-ton open-hearth furnace at its new Trenton plant in 1883, and this remained the only open-hearth operation in Canada until 1900. The chemical process within these new furnaces had replaced the puddler's craft, but human expertise and muscle were still straining to produce the steel. An open-hearth "melter" and his crew charged the furnaces by placing the scrap, iron ore, and chemicals on long, flat-ended iron rods known as "peels" and extending them manually into the furnace. According to one observer, this was "one of the hardest and hottest jobs known in the steel business," and an open-hearth worker "was considered old at 40 years."[10]

From the puddling or open-hearth departments, the metal was carried to the second stage in the two-step process designed to escape the necessity of hammering out wrought iron – the rolling mill. By this method, heated iron blocks from the puddling or open-hearth furnaces could be passed back and forth between two heavy, grooved rolls, rotated by the power of a steam engine, until the iron had been sufficiently flattened and stretched. In this process, the iron went first through "mucking" or "roughing" rolls, then back to a reheating furnace in bundles of bars to be fused together, and finally back to a "finishing" roll, which turned out bars of so-called "merchant" iron ready to be sold. The first rolling mill at a Canadian ironworks began operation at the Londonderry site in 1860 and several others appeared in other primary producers' enterprises over the next four decades. Simultaneously, of course, railway corporations and metal-trade manufacturers were opening independent rolling mills in Hamilton, Toronto, Montreal, Saint John, and elsewhere.[11]

The painter William Armstrong immortalized one of these, the Toronto Rolling Mills, in 1864.[12] His painting reveals the inevitable billowing smoke and glowing iron beams, but also the reliance on human muscle power for shifting, heaving, thrusting, pulling, and otherwise handling the iron bars throughout the production process. Originally, since rolling mills had only two rolls, the iron had to be passed back and forth over the top of the machinery. A reversing engine was introduced, but North American mill owners more commonly added a third roll above so that the iron would simply be lifted and passed through for its second squeezing. In any case, the process still relied on the muscle and savvy of the "heaters," "catchers," "roughers," and, above all, the lead hands, the "rollers" – all of them strong-backed men who used simple hooks and tongs to snatch the hot ingots or bars from reheating furnaces and thrust them into the successive sets of rolls, and who judged when the iron had reached the right size and shape. These were clearly labour-intensive operations. A visitor to the London-derry rolling mill in 1881 described how "the men are like bees in a hive hurrying to and fro," and an executive of the Ontario Rolling Mills in the same period later recalled:

> It is a singular fact that in the industries of that day almost no labor-saving machinery was used. All the lifting was done directly by hand labor and so cheap was this labor that it was really more economical to employ direct human effort than machinery. Even cranes, derricks, tackle and all the various simple labor-saving devices were rarely used in any factories at that time.[13]

The Canadian iron industry had thus adapted to the new age of industrial capitalism in the second half of the nineteenth century with considerable technological innovation. Special ovens had been developed for preparing charcoal and coke. Steam power had made blast furnace work somewhat faster and more efficient. Refining pig iron into wrought iron or steel was a new process involving, first, puddling furnaces and then, in one case, open-hearth furnaces. And powerful, steam-driven rolling mills had replaced the manual procedures of the blacksmith. Overall, the scale of production had increased considerably. Yet the impact of this new technology had not been to eliminate much manual labour. The machines remained relatively simple and disconnected and required plenty of human intervention to feed and run them and to monitor their output. New groups of skilled workers had emerged, notably around the puddling furnaces and the rolling mills, and huge numbers of

Rolling-mill hands at the Hamilton Steel and Iron Company, 1906. (McMaster University Labour Studies Collection)

labourers still lifted, heaved, carried, or wheeled through the smoky gloom of these large plants.

The Second Industrial Revolution

The blowing in of Disco's first blast furnace in December, 1900, marked a dramatic turning point in Canadian iron and steel production. The huge new steel plants of the early twentieth century housed much more sophisticated technology than had operated only a few years earlier.

Typically the raw materials now arrived in ships or railway cars. Unloading and moving them into the plant no longer required either human or animal muscle. One man would activate each of the steam-driven or electrically powered machines that hoisted coal or ore out of ships' holds and dumped it into waiting railway hopper cars (or smaller transfer cars). In Sydney four of these lifting devices could handle eighty tons per hour, and by 1926 they could unload 10,000 tons of ore in one night. The drivers of the trains would then move the stock over to bins by the coke ovens or the blast furnaces for immediate use, or else up onto trestles over the storage yards. In both cases, the material would simply be dropped out the bottom or side of the cars. Out in the yards, operators of huge electric travelling cranes known as "bridges" would scoop up the material when needed and drop it into railway cars for its trip to the bins. "From the time the coal leaves the mine until it enters the blast furnace stacks as coke it is untouched by shovel or fork," a journalist reported in 1924. Machines had replaced most of the straining muscles in the old nineteenth-century routines.[14]

Preparing coke for the blast furnace was similarly mechanized. The first step had been to install bins and charging machines above the ovens and other electrically powered machines for levelling the coal and pushing the coke out. At Sydney cooling and heaving the coke by hand continued for a few years until conveyor belts, chutes, special cars, and "quenching" apparatus replaced workers' manual labour, as they had at Scotia, Algoma, and Stelco. Even the opening of the oven doors was electrically powered, thus "doing away with all hand labor," Algoma's coke oven superintendent explained in 1919. "Half a dozen men now make a larger amount of coke than was formerly made by half-a-hundred men," a business journalist reported in 1924. This was a far cry from the farmers' burning logs in the backwoods of British North American colonies.[15]

So, too, was the process of charging the blast furnace. The raw materials that had been stored in elevated bins behind the furnaces were now released into small, electrically driven "scale cars," which automatically measured the weight. The drivers of these vehicles then moved them along railway tracks and emptied the contents into a lifting device beside the furnace known as a "skip hoist," a kind of conveyor belt with attached metal boxes that hoisted and dumped the stock into the top of the furnace with no manual assistance. At the front door of the furnace, small quantities of iron might still be cast in sand, but more commonly crane operators caught the molten pig iron in huge buckets (always known in the industry as "ladles") and swept it away for immediate use in the steelmaking departments, or else emptied the hot metal into "pig-casting machines" – conveyor belts of steel moulds that cooled the metal quickly, dropped the pig iron into a railway car, and then returned underneath to receive another load of hot metal. Gone from all this work were most of the shovels, wheelbarrows, and small armies of brawny labourers, and in their places were a few handfuls of men who manipulated gears. "Under the old style it took 150 men per 24 hours to operate a 200-ton furnace and the output was 1.33 tons per man turn," a Canadian trade publication reported in 1924. "Under the modern style it takes only 60 men per 24 hours to operate a 550 ton furnace and the output is 9.17 per man turn."[16]

The ladles of molten pig iron that travelled along at the end of the powerful new electric cranes arrived in steelmaking departments that dwarfed the old puddling operations (and even Scotia's tiny open-hearth facilities). Algoma's steel plant revealed the greatest novelty, since the corporation installed technology developed in Britain and widely used in the United States over the previous quarter century, but never before seen in Canada. A British inventor by the name of Henry Bessemer had discovered that blowing air into and through the molten iron to oxidize the impurities would create enough heat to keep the metal liquid. The speed of this process, and the avoidance of skilled labour and expensive fuel costs, had been immediately attractive. At the beginning, three to five tons of steel could be produced in ten or twenty minutes.[17] In the United States, ironmasters who produced steel rails had made the Bessemer method the key to the success of their huge new mills by the 1880s, and Clergue therefore decided to copy their example. A local reporter watched the first steel cast in the Sault in 1902 in wide-eyed wonderment. The molten pig iron, delivered in a huge ladle, was poured into one of the two large, pear-shaped, five-ton

The open-hearth department at Algoma Steel in 1918. (Public Archives of Canada, PA 29355)

converters, "which looked like a beer bottle, on which the neck had been put a little sideways and then broken off short." The sound and light show began at once: "When the blowing engine was turned on the sparks came out of the crooked neck in a steady flame that filled the whole south end of the building. . . . In a little while the loudly roaring fire was too white to look at. Then the 'blower' knew the 'cook' in the vessel was 'done' – was cleansed enough to pour." A shift of another lever tipped the converter to let the steel stream out into a waiting ladle, which a crane operator lifted and swung over a row of ingot moulds. A worker then released the steel from the ladle into the moulds. The whole process had taken little more than half an hour. Later the ingots would be stripped of their mould casings by another specially designed crane.[18]

As Algoma soon discovered, however, the problem with the Bessemer process was that it required pig iron made from ore without much phosphorous, which was increasingly scarce in North America.[19] By the turn of the century, a great conversion to open-hearth production was under way across the continent. That process had seen some significant changes, however. The open-hearth departments that opened in the main Canadian steelmaking centres between 1900 and 1907 incorporated numerous "labour-saving devices" to eliminate much of the manual labour needed at Scotia's first plant in the 1880s. The skilled "melters" still had considerable discretion in determining the quality and timing of the brew, and at regular intervals their helpers had to face the gates of hell with shovelfuls of chemicals. But machinery had replaced most other manual labour. Men now drove small vehicles mounted on railway tracks for thrusting iron and scrap into the furnaces. Others used powerful electrical cranes to pick up ladles of molten steel and to fill ingot moulds sitting upright on railway cars ("teeming the ingots," in the jargon of the trade). Others handled the "stripping cranes" for removing the moulds, while another crane operator carried the ingots off to so-called "soaking pits," deep ovens that heated them to a uniform temperature for their trip through the rolling mills. A few key workers were still responsible for quality control, but, in contrast to the slow, arduous, manual routines of the old puddlers and bushelers, many more men in the open-hearth department were operating or helping to maintain machinery, which handled the tasks of large-scale steel production quickly and smoothly.[20]

The steel ingots that left the Bessemer and open-hearth depart-

A crane operator used a giant "ladle" to catch liquid steel pouring out the back of an open-hearth furnace at Stelco's Hamilton plant in 1918. (Public Archives of Canada, PA 24646)

ments then entered perhaps the most startlingly transformed technical arena of the whole steel plant. Many of the new rolling mills were machines of monstrous proportions never before seen in Canadian iron and steel production. In these new operations, a crane operator would fetch a reheated ingot from the "soaking pit" and place it on a bed of electrically driven rollers. From here, men perched high up on platforms known as "pulpits" controlled the complex machinery by pulling levers and pressing buttons, sending the hot steel block back and forth through the rolls. Each time through, the flip of one lever would raise or lower the large table on which the ingot rested, while another would activate mechanical arms called "manipulators" for turning it over. The steel "bloom" that emerged from this process would then be sent along more automatic conveyors to giant shears, where men would cut it into appropriate lengths, and then on to smaller sets of rolls to produce billets, rails, and so on. The newer installations often had "continuous" furnaces and mills; that is, the steel would pass in one direction through them rather than reversing direction frequently. As the *Canadian Foundryman* noted after surveying Stelco's new continuous, electrically driven operation in 1913: "Modern rolling mills are really automatic machines on a large scale, one machine sometimes covering an acre or more of ground, and operated by a few men almost entirely without hard muscular labour."[21]

As the steel passed on to the progressively smaller finishing mills, however, the reliance on human labour increased. Most of the older Canadian rolling-mill operations were still running on the kind of labour process that had predominated before the turn of the century. They still produced relatively small batches of varied products for the limited Canadian market and consequently could not afford the expensive investment in highly specialized, highly mechanized processes. In 1910 a supervisor at Stelco, which brought a number of these older operations into its merger that year, told a parliamentary committee: "We would consider we would have a very good order if we ran a day on it." A federal conciliation board and a royal commission heard similar evidence of labour-intensive technology in parts of Scotia's rolling-mill operations in 1915 and 1919. Even in the newer plants, the work process on some of the smaller rolling mills, like the merchant mills, involved heavy manual labour. For some rolling-mill products, moreover, new, labour-saving technology had simply not been developed by World War I. Stelco's new sheet-metal mill installed in 1918 was the best example, with its heavy demand on skill and muscle.[22]

The steelmaking corporations would continue to attempt to modernize these older, smaller rolling-mill plants, in some cases right down to the 1940s, by increasing the degree of mechanization. Yet the constraints of markets and long-term indebtedness could make such innovations extremely difficult. It is important to bear in mind, therefore, the unevenness of the transformation in Canadian rolling mills and the co-existence of the hand-operated finishing mills and the mechanized giants well into the early twentieth century.

The first steelmaking corporations, of course, had done more than simply change individual work processes. They had brought most of them together in huge sprawling complexes that tightly linked each phase of production with the others. Individual departments and processes within departments nonetheless retained some distinctive rhythms and occupational requirements. Some processes were normally continuous, notably the coke ovens, blast furnaces, and open-hearth, while others, especially the smaller rolling mills, were based on batch production. Some, like blast furnaces and rolling mills, required steady, routinized feeding of furnaces or machinery, while others required more erratic bursts of frantic exertion, especially tapping furnaces. There was, in short, a great variety of work in a steel plant that did not necessarily lead to a common occupational experience for the industry's workers.

In many ways, the sum of all the technological innovations was greater than its parts. Besides the changes in individual stages of production, it was the thorough integration of the plants that was so remarkable. Every plant was a maze of tracks for numerous railways, cranes, and conveyers. Raw materials moved along these to coke ovens and blast furnaces; liquid pig iron was swept off to the open-hearths at the end of giant cranes; steel ingots were shunted off to the rolling mills, where cranes and conveyors carried the steel forward. In contrast to past practices, there was far less remelting and reheating of the metal as it moved through the plants. Early twentieth-century steelmaking did not involve an assembly line, but there was definitely an integrated flow-through. For the most part, moreover, the plants ran continuously throughout the year, rather than on the seasonal basis that had characterized much nineteenth-century production. In fact, several departments ran non-stop twenty-four hours a day and seven days a week. Mechanization therefore brought not only greater volume of production from the new facilities, but also greater speed and intensity and, for the workers, greater pressure to keep up.

In the opening years of this century, the whole steelmaking labour process, from iron ore to finished steel, thus took on a characteristic shape that would not change fundamentally before the 1950s.[23] Technological change, however, was not a single event. A process of continual renewal and updating would be necessary if Canadian firms were to keep abreast of the latest, most productive technology. All the Canadian plants began with some significant pockets of technical backwardness, and by the 1920s Algoma and Disco were falling further and further behind in the process of upgrading and streamlining. In contrast, Stelco used the 1920s to refurbish most of its plants with new labour-saving machinery and more flexible electrical power.[24]

Most contemporary reports nonetheless emphasized how the Canadian steel industry, like its American counterpart, had been transformed by the machine. The *Journal of Commerce*'s editor, A.R.R. Jones, applied the label "gigantic automaton" to the "typical all-round Canadian steel plant" that he visited (clearly Stelco's Hamilton plant), in which "the labor in every branch of the industry consists mainly in the supervision and maintenance of machinery."[25] What was missing from these glowing descriptions of mechanization was what the steelworkers who worked with this new technology every day had discovered. It was remarkably dangerous. The intense heat from furnaces alone could inspire fear, but the showers of sparks from ladles of molten metal could actually sear the flesh of nearby workers. Photographs of steel production from these early years indicate how little protective clothing or equipment was used. The noise could be similarly fearsome. One man described how on his first day in Algoma's plant he could not hear a train that he suddenly discovered was passing inches from his back. Not surprisingly, an American writer found that many of the steelworkers he met had hearing problems. If the men were not dodging locomotives or machines whose tracks criss-crossed the plants, they were scampering out of the way of ladles, moulds, and great hunks of glowing steel that soared through the air at the end of giant cranes. For one Stelco worker, the first day on the job was "like entering another world." For another man at Algoma, this mechanized work world seemed like "organized confusion." Stress would inevitably become a new occupational hazard in such a fearsome workplace.[26]

Under these circumstances, the accident rate was deplorably high. Industrial accidents were not reported with any consistency before 1910, but in Ontario, in 1912, at least 388 Algoma employees and

113 at Stelco suffered occupational injuries that required at least a week off work. The same year the accident rate at Disco was so high that Nova Scotia's factory inspector decided to conduct a special investigation. He examined a sample of 100 serious accidents and fatalities in a three-month period and concluded that 89 of them "were due to the ordinary risk of the occupation, and could not be prevented by any known guard or protection." Four years later, the last detailed provincial statistics available included 473 at Stelco (almost one worker in six), 488 at Algoma, and 345 at Disco. Thirteen of all these were fatal. Between 1916 and 1920 the new Workmen's Compensation Board in Ontario recorded eighty-two deaths, 410 permanently disabling accidents, and over 9,000 less serious injuries in the province's iron and steel plants. Small wonder an Ontario Safety League official would later describe a Canadian steel plant as "the worst industrial slaughter house in the whole of Canada."[27]

That ever-present danger could breed fear in some steelworkers, but, for many, it posed the challenge to prove their manhood. Among the committed men in the industry, there was (and still is) a masculine pride in their ability to face the gates of hell.

Technology and Skill

What had prompted this wave of mechanization in the steel industry? Most commentators on the complex new machinery in Canada's steel plants praised, above all, its "labour-saving" qualities. This popular phrase did not mean saving workers' sweat but, rather, eliminating manual labour and, wherever possible, workers themselves. For manufacturers in such a traditionally labour-intensive industry, who were struggling to meet foreign competition, a central concern was inevitably cutting labour costs. And a scarcity of labour, too, often pushed up those costs.

Skilled ironworkers had always been rare commodities in the Canadian labour market and usually had to be imported. An early nineteenth-century ironmaster had voiced a complaint familiar to iron men throughout the century when he described skilled ironworkers as "the very worst sort of men to manage" as a result of their labour-market position: "Not one of a hundred of them but will take advantage of his master in his power," he wrote. "If I have just the number of hands for the works, every one of them will know that I cannot do without everyone of them; therefore everyone of them will be my master."[28]

Before 1900, however, the unskilled labour needed for the bull work had seldom been a problem. As the Ontario Rolling Mills official had admitted, an abundance of cheap unskilled help discouraged entrepreneurs from installing expensive new equipment for handling materials and products. But in the economic boom that began at the turn of the century, steelmasters often encountered difficulties in obtaining unskilled workers, and at various points before 1920 they faced shortages in the local labour markets (especially at "Harvest Excursion" time in the early fall) that could hamper production and force up wage rates.[29] The many hard, backbreaking unskilled jobs in the older mode of iron and steel production were never popular if other work was available. "Workmen find it bad enough to be forced to handle frozen pig and scrap iron in winter," the *Canadian Foundryman* lamented, "but when the summer heat comes beating down the men become inefficient and discontented. Many of them leave."[30] Of course, the large new steel mills in the small, more isolated communities of Sydney and Sault Ste. Marie could probably never have relied on local labour supplies for their vast operations. The logic of labour markets, then, suggested a managerial policy of substituting machines for men.[31]

So, too, did a concern with large-scale output. Once a manufacturer could be assured of a standardized demand for a high volume of output, investment in mass-production technology made good business sense (this would become Henry Ford's secret to success in the American auto industry). In the early years of the Canadian steel industry, the most mechanized plants were those that anticipated a high demand for a standardized product, namely, steel rails, while the slowest to change were those parts of the industry that could not rely on this kind of large market, especially the rolling-mill departments at Scotia and Stelco. Of course, an initial investment in expensive machinery encouraged steelmasters to keep it operating at peak efficiency and, as a result, to install still more mechanical devices for speeding the flow of production through their newly integrated plants.

The steelmaking corporations that emerged in Canada at the turn of the century benefited from three or four decades of experimentation in the United States aimed at eliminating bottlenecks in production created by slow, manual routines. By the 1890s these experiments had resulted in high-volume mass-production in the major American plants. With the partial exception of Scotia, which was somewhat more eclectic in its technological derivation,[32] all the

Canadian steel companies imported the machinery for their plants from the United States and continued to watch the trends in American production techniques. The leading Canadian steel men were also regular participants in the annual meetings of the American Iron and Steel Institute, presided over by US Steel's Judge Elbert H. Gary.[33] The first annual report from the directors of Algoma Steel, which had purchased the equipment of a steel plant in Danville, Pennsylvania, and moved it to Sault Ste. Marie, pointed out the advantages of this American technology: "The arrangement of the plant is such that material can be handled at a minimum labor cost, and an unusually large output per man is thus obtainable."[34] A decade later a Nova Scotia royal commission observed that Disco's work force had been "enormously reduced" in order to "eliminate needless cost wherever possible." By the 1920s a business journalist could conclude that "the products could not be turned out at anything even approaching their present low price if human labour had to be utilized to do what is now done by mechanism."[35]

Certainly the available statistics on wages as a percentage of the value of production suggest a steady decline. Ontario's Royal Commission on Mineral Resources calculated in 1890 that labour costs in producing pig iron could vary between 9.3 and 25.8 per cent of the total production costs.[36] In contrast, federal Mines Branch figures for the 1900-1920 period indicate that, while the new steelmaking corporations often had trouble keeping these costs down, blast-furnace wages were generally in the much lower range of 5 to 10 per cent. Stelco's percentage alone dropped from 16.4 per cent in 1904 to 4.2 per cent in 1920. Similarly, statistics on output per man in early twentieth-century Canadian steel plants reveal great increases in productivity. Between 1905 and 1920, each blast-furnace worker's output rose 25 per cent at Disco, 72 per cent at Scotia, 142 per cent at Algoma, and 198 per cent at Stelco. In 1920 each open-hearth worker was turning out 12 per cent more at Disco, 25 per cent more at Scotia, 119 per cent more at Algoma, and 166 per cent more at Stelco than in 1907.[37]

The question remains: what kind of labour was eliminated? What was the impact of technological change on skill requirements in the steelmaking labour process? Before considering in more detail the specific skills of the work force in an early twentieth-century steel plant, however, we should review the main arguments of a growing body of historical and social-scientific literature on the question of skill and deskilling. In much of this writing, skill is correctly perceived to be a complex blend of technical competence and "social construction" (that is, subjective assertions of what a "skill" involves). Skilled workers combine manual dexterity and conceptual abilities and usually exercise some degree of autonomy and discretion on the job (within the inevitable constraints of a wage relationship). Quite often, however, their position is buttressed by some kind of social sanction, either organizational, in the form of union controls, or ideological, as in the assertion of the craftsman's "manhood" and "respectability," or in the exalted idea of "craftsmanship" itself. Loggers and coal miners needed a technical competence on the job that was seldom granted the status of skill in large part because of the roughness of the work. The traditional failure to recognize so many of women's skills, like needlework or typing, has also underlined the importance of social constructions of skill within a patriarchal capitalist society. Most often, it should be added, the social construction of a skill either embellishes or undervalues its real technical content; few workers can use this element of subjectivity to protect technically unskilled work.[38]

A substantial body of writing on the labour process, much of which takes its cue from Harry Braverman's *Labor and Monopoly Capital*, has focused on the craftsman in late nineteenth-century North America as the last embodiment of skill and has identified the final destruction of skill in the obliteration of that romantic figure through the consolidation of mass production.[39] Braverman, like Marx,[40] places great emphasis on the role of new technology in this process of wrenching control away from skilled workers, and, in his eyes, the impact of mechanization was to reduce drastically the skill levels of the industry in question. The degradation of labour, and of the worker, was the inevitable outcome: "The 'progress' of capitalism seems only to deepen the gulf between worker and machine and to subordinate the worker ever more decisively to the yoke of the machine."[41] A similarly influential article by Katherine Stone makes the same argument for the American steel industry. Stone stresses that the new steelmaking routines required workers only "to operate the machines, to feed them and tend them, to start them and stop them." Stone also credits F.W. Taylor, the prophet of "scientific" management, with introducing new managerial practices in the industry for breaking down skill, removing from workers the conceptual function in the production process, and placing it in the hands of front-office planners. For Stone, the destruction of workers' shop-floor discretion was complete: "Steelworkers no longer make any decisions about the process of making steel."[42]

The theoretical assault on this analysis has been telling. Several writers, especially labour historians, have challenged the emphasis of Braverman and others on one-dimensional degradation through technological change. They have stressed in particular working-class resistance to managerial innovations, both in informal work groups and through new unions.[43] Those studying work relations in Britain have also been the most astute in noting the unanticipated outcomes of the struggles for workplace controls, which could leave workers with more shop-floor power than owners and managers would have liked or than theorists would have predicted.[44] For these writers, "deskilling" is a far too simple and misleading notion, since they have found evidence that old skills survived and new skills emerged in the process of industrial transformation. Certainly the assumption that mechanization inevitably means deskilling has been far too glib. What has too often been lacking from these discussions is a careful examination of the real impact of specific machines in specific industries and of the new meaning of skill amidst the new technology. Recent discussions have argued that there are "tacit" skills that emerge in these new settings,[45] and even that there has been an "intellectualization" of skill in learning to handle intricate technology.[46] A closer examination of the evolution of skill requirements in the early twentieth-century Canadian steel industry will reveal that Stone's analysis has missed this complexity and seriously misjudged the consequences.

Certainly a large percentage of unskilled bull work in steel production was gone – lifting, heaving, carrying, wheeling, and so on – though not all of it, of course. Despite the technomania of business and trade journals, unskilled labour could certainly still be found in Canadian steel plants in considerable numbers in the early twentieth century. Out in the yards, men often moved scrap or other materials by hand. Others shovelled coal for boilers or stoves or carted bricks for relining furnaces. Moreover, steelmaking was an untidy process with huge messy spills to be cleaned up by gangs of men at the base of the furnaces. From time to time, labourers also had to crawl inside furnaces to clean out old bricks or slag, or under furnaces and stoves to shovel out ashes. Many of these men would work in a distinct administrative unit known as "General Yard." Overall, however, the need for such labour had declined to the extent that a Disco official could inform a royal commission in 1923 that only 20 per cent of the corporation's work force was "common labour."[47] Evidently, cranes and other conveyors had proven stronger, faster, and more reliable than labourers prone to

Travelling cranes with electric magnets made moving scrap from Dofasco's storage yards during World War I much faster than it would have been with manual labour. (Public Archives of Canada, PA 24528)

working at their own pace, striking, or leaving in search of less taxing jobs.

Steel plant owners had been just as concerned with problems of skilled labour. Here they made some strides but also encountered some intractable difficulties. Whatever machines were used, iron and steel production required judgement based on a familiarity with metallurgical chemistry that could only reside in a human mind. At the core of a successful nineteenth-century iron works had been a furnace-keeper who understood the properties of iron, could concoct just the right mixture of raw materials to produce first-rate pig iron, and could supervise the small furnace crew. Few of these skilled men had much formal training. They learned a kind of rule-of-thumb, trial-and-error metallurgical science on the job from older skilled workers (there was no well-established apprenticeship system among these workers).[48]

The major assault on their importance in the production process came from the new university-trained chemists that each of the new steel companies hired to run research laboratories in their plants. Scotia's Scottish furnace-keeper had little time for the American chemist he found at the Ferrona works in the 1890s. "Instead of making use of the Laboratory," a company official later wrote, "his experienced eye would judge the operation of the furnace by the color of the gas and slag." Yet this worker and his counterparts in the other Canadian plants had to come to terms with this new group of professionals that their employers had inserted into the labour process. "No up-to-date plant is equipped for work without a well fitted laboratory for speedy work, all supplies and output being bought and sold on analysis," a Canadian writer explained in 1896. "Iron and steel works' chemists have during the last few years converted the use and manufacture of iron and steel from the old 'hit-or-miss' methods to science." A surviving letterbook from Disco's first chief chemist indicates that his staff of eighteen-twenty was kept busy day and night in the early 1900s analysing samples brought in from the furnaces by errand boys. Francis Clergue was proud of similar facilities he had set up in Sault Ste. Marie to work at applying the fruits of scientific research to the practical problems of production.

Yet the know-how acquired in and around the furnace itself would give the blue-collar furnace-keeper an edge over the white-collar professional for years to come. Scotia's general manager explained this problem in a letter to his Scottish labour agent in 1910. He was disappointed that the young chemist recently sent over had not had enough "practical knowledge of the working of furnaces"; the man needed "an inclination to spend more of his time about the furnaces and watch and assist in their working." The workers themselves could get frustrated at the inexperience of the college-trained men. In the fall of 1903 a newspaper story described some Sydney blast-furnace workers' complaints about having to make up for the professionals' incompetence. "It is not that they deprecate college training," the paper noted, "but they hold that it would pay the company better to take the men at hand, the self-made mechanics who had gathered their knowledge from practice, instead of taking men without experience." A Canadian metal-trade journal similarly concluded that blast-furnace operation demanded shop-floor experience and skill:

Operating a [blast] furnace is a simple and routine proposition when all is going well, but furnaces are known to their keepers as members of the female sex, and they have all the beauties and uncertainties of that sex, at times they do not go well. The iron notch cannot be opened, and cinder reaches the tuyeres, or the stock sticks, scaffolds and freezes, to say nothing of blow-outs and gas leaks. All these things call for a high degree of skill and intelligence on the part of an operator, for the false move on the part of someone may endanger a life and property.[49]

A similar dilemma faced the new corporations in the departments where pig iron was turned into steel. The major casualty in the technological transformation of the late nineteenth century had been the iron puddler, who had built up not only the muscle to "puddle" a furnace full of pasty iron, but also the savvy to control the quality of the product. The new Bessemer and open-hearth furnaces boosted the output of steel enormously, but determining the right mix of additives and judging when the molten brew was ready were new skills similar to those of the old-time puddler. The "blower" who headed the work team at the Bessemer converters and the man at the head of each open-hearth crew, known as a "melter" (or "first helper"), were seasoned, shop-floor experts on whose judgement the steel companies relied for the quality of their primary steel. The first melter employed at Nova Scotia Steel in the 1880s had to be imported from England, and the new plants in Hamilton, Sydney, and Sault Ste. Marie brought theirs from the United States. Most often, replacements then learned their skills on the job in Canadian plants. A Sydney steelworker who began work in Disco's open-hearth department in 1922 recalled the control

"Melters" were largely responsible for quality control in Stelco's open-hearth department. (Public Archives of Canada, PA 17740)

over the work process that these men and their crews enjoyed. Not only did the melter decide on the number of boxes of ore and scrap to be added before the liquid pig iron arrived, but he would have to take a test of the brew. To do this, he would scoop out a sample with a small, long-handled ladle, cool it quickly, and break it to assess the grain of the metal for carbon content. The foreman's role in the department was simply to indicate the kind of steel to be made (soft or hard, depending on the carbon content); "and he'd be gone and you wouldn't see him anymore till the next heat you'd be ready to tap. Everything depended on the men on the furnace. It was in their hands." This was another arena where the corporations' professional chemists intervened, but, from this old steelworker's account, it appears that tests would be sent to the laboratory only once the melter and his crew were sure the brew was ready. The industry's trade journal had to admit in the 1920s that the melter "is usually a regular wizard in his exact knowledge of how the refining process may be facilitated from time to time."[50]

The third arena of persistent skill content was the rolling-mills department. In the second half of the nineteenth century, the black-smith's sinewy arms had given way to steam-powered rolling mills, but mechanization had not produced a work force of unskilled machine-tenders. The new groups of workers who manipulated the iron and steel back and forth through the rolls set their own pace and judged the quality of the product themselves under the direction of a head roller. Nova Scotia Steel had to import its first rollers from Saint John, and in 1888 Londonderry's rolling-mill superintendent told a royal commission that the absence of a roller could curtail production: "We have no spare men with the particular training of rolling, and the roll would have to stand idle."[51] The massive new installations in the early twentieth-century plants eliminated much of this direct human element in the "blooming" mills that reduced the steel ingot in size and in the rail and billet mills that further refined it. As we have seen, these massive new machines had many more automatic features, which were manipulated by machine operators controlling a nest of levers. The rollers nonetheless had to monitor the thickness of the steel on gauges and to adjust their delicate levers accordingly. In some of the further finishing processes, moreover, the shop-floor judgement of the rollers remained critical. A steelworker who started on Disco's bar mill in 1925 remembered many years later how the "finisher," the last man to pass the bar through the mill, monitored its thickness with a stick and adjusted the rolls accordingly, "so that the finished prod-

Like these workers on Algoma's merchant mill in 1918, the men who worked on the smaller, more labour-intensive rolling mills still exercised considerable shop-floor discretion. (Public Archives of Canada, PA 29348)

As late as the 1930s, these men on Stelco's sheet mill were still a highly skilled group. (McMaster University Labour Studies Collection)

uct will be what the customer wants within seven thousandths." An inspector at the end of the line would reject any faulty bars, but, as this man recalled, "we've rolled many, many days without one rejection." The same story was told in other plants. "We knew when we were handling that steel what its potentials were, what we could and couldn't do with it," a Stelco sheet-mill worker explained many years later. "We had to do it all with our eyes and judgement, mainly judgement. There were no gauges like today." Not only did these skilled rolling-mill hands have considerable discretion on the job, they worked with the same light supervision as the open-hearth men. "The foreman would come around at the start of the shift, mark their clock numbers on a box of cigarettes, and you wouldn't see the foreman again until the end of the shift," the same worker explained.[52]

An important new skill found in all departments was operating the huge overhead travelling cranes. Incompetence in transporting tons of molten iron and steel could result in great expense to the company and danger to the crane operator's workmates. One former Sydney steelworker explained that as an operator or a "chaser"

The new groups of skilled workers needed in Canadian steel plants included crane operators, like this one filling ingot moulds with molten steel at Disco's Sydney plant in 1903. (Canadian Mining Review, February, 1903)

(the man who followed the crane on the ground), "if you weren't skilled you could make one hell of a mess." These skills parallelled those needed by the locomotive engineers who moved materials along the miles of tracks in and around the plants.[53]

So, while most of the old-time craftsmen of nineteenth-century ironmaking – the puddlers and the blacksmiths – were gone, all skill did not disappear within the "gigantic automatons," and managers recognized their reliance on such men. Stelco's open-hearth superintendent told a parliamentary committee in 1910: "You cannot go and round up the skilled men and pick them up on the street corner. Take our melters, rollers and first helpers, they are skilled men and the next man to one of these cannot take his place . . . the same with the men at the ladles. . . ."[54] He might also have noted the considerable numbers of machinists, moulders, blacksmiths, stationary engineers, electricians, and other craftsmen who were now needed in much greater numbers to maintain the complex new machinery. Disco had 400 of them by 1910 (in a work force of some 4,000), and by 1923 its "Mechanical Department" that grouped these workers had 891 men (out of a total of 3,862). The firm continued to maintain a small apprenticeship system for training some of these valuable workers.[55] Not surprisingly, many of these skilled men in Canada's steel mills took the same pride in their role as producers that the old-time craftsmen expressed. "When I worked there," a retired open-hearth worker insisted, "men were proud to work there, and they took pride in their work, I'll tell you that. . . . It's the men on the furnace – that's who makes the steel."[56]

Without a complete payroll list for any of the plants, it is difficult to determine the precise proportion of the new steelmaking work force that might be considered skilled, but there was undoubtedly a considerable number of steelworkers whose knowledge and judgement were crucial to the production process, especially in the open-hearth and rolling-mill departments. At the same time, however, it is clear that, relative to the volume of output, far fewer of them were needed than in the nineteenth-century mills. Nine puddlers at Londonderry could produce only twenty-five tons a day, whereas by the 1920s the three-man crew on each open-hearth furnace could produce up to four times that amount in one eight-hour heat.[57] They were consequently less expensive for the steel companies, since the cost of their labour added less value to each ton produced.

The new mechanized processes in Canada's steel plants were thus reducing the demand for labour at both ends of the occupational hierarchy. But they were also opening up a much larger category of new jobs in between – the so-called "semi-skilled" jobs of operating the new machines. This major new occupational group within the twentieth-century working class has perplexed social analysts for some time. The optimists who celebrated the successes of capitalist industrialization have characterized the appearance of these workers in industry as a process of upgrading the less skilled into better jobs. Hard-nosed critics of these developments, however, have pointed to the extremely low training requirements for such jobs and stressed the mindless repetition and grinding degradation of the work.[58] Neither is an adequate perspective on what happened in the steel industry (and probably in several other mass-production industries as well), since both miss crucial dimensions of semi-skilled work. We need to look more carefully at the jobs of these machine-operators.

A special report on the American steel industry just before World War I caught the ambiguity of this work force when it noted that the semi-skilled were "workmen who had been taught to perform relatively complex functions, such as the operation of cranes and other mechanical appliances, but who possess little or no general mechanical or metallurgical knowledge. . . ."[59] These men did work that demanded more responsibility and competence than labourers, but less general knowledge than craftsmen. Normally the time needed to learn a job tends to attract most attention in determining the technical content of a skilled job, not least because that criterion is directly related to how easily a worker can be replaced by another from the reserve army of the unemployed. In an address to the American Iron and Steel Institute in 1919, Stelco's president proudly announced that in four and a half years of operating the corporation's new electrically driven blooming mill, less than an hour and a half had been necessary for "breaking in new men to operate the motor for the mill."[60] The implication, of course, was that replacements could have been trained as quickly and easily.

Yet this industrialist neglected to comment on the apparent fact that these trained men stuck to their jobs, and that their accumulated experience with the intricacies of the complex machinery probably helped to maintain peak efficiency in the plant during wartime. Certainly other steel-company managers emphasized the importance of experience in semi-skilled work. Disco, Algoma, and Stelco all had to import large numbers of men familiar with steelmaking routines, mostly Americans, to run the new machinery they installed at the turn of the century. Typically, the corporations' supervisors (also American) reached out to the network of men whose competence they knew from working with them in American mills. "As we expected," Disco's blast-furnace superintendent wrote to a friend he was attempting to lure north, "the labor here is poor, from our point of view, not having any experience in this work, but we are getting men from the States, such as Keepers, Helpers and Mechanics, and doing our best to get an organization that would resemble our works in the States." He also complained to other company officials that "It seriously interferes with our work to be breaking in new men all the time," and that "the crews are changed so often that we get green men, unfamiliar with the work, and confusion and mistakes arise as a consequence." A few months later, his successor begged the general manager to get "some competent man" to run the pig-casting machine, which, as we have seen, was a relatively simple conveyer belt. Toward the end of the year, he reported similar problems with the machinery for weighing stock to be dumped into the furnace and requested "some expert scale man, who is capable of keeping these cars in condition."[61]

Many other jobs around a steel mill demanded this kind of familiarity with the machinery and the properties of the metal being handled. A reporter visiting the Algoma rail mill in 1902 learned that the man who chopped up the steel blooms with giant shears had to judge the quality of steel bars approaching him from the rolls and to reject any unfit to be sent on to the rail mill. Four years later a technical article on Disco's rail mill noted the similar care needed in shearing off the ends of the rails to take account of shrinkage when hot: "A skilled man handling the stops can, from experience, judge the precise temperature of the rail bar when cutting and hence places the rail stops so accurately that all the rails are found to be the required length when cool." A little further along, the writer observed, the rails passed through a "cambering machine," which a skilled operator could adjust "so as to get an almost straight rail when cooled off on the hot bed." When a controversy blew up in Sault Ste. Marie in 1907 over the large number of Americans filling these positions, an Algoma official explained:

> Rolling mill work is somewhat different to other manufacturing business as incompetent men can lose money for a company very fast. One mismove of the hand of a workman or operator of

Controlling the levers on Nova Scotia Steel's "cogging" (blooming) mill in 1916 demanded careful attention. (Beaton Institute, College of Cape Breton)

So-called "semi-skilled" workers had to handle some complex machinery, such as this charging machine in front of Dofasco's furnaces around World War I. (Public Archives of Canada, PA 24525)

the machine can spoil $200 worth of steel in a couple of seconds without counting the delay or the damage done the machinery. So any fair-minded man can see that a mill superintendent has to be very careful as to who he employs.[62]

These examples could easily be multiplied, but the point is evident: regular and efficient production in Canada's steel plants depended on the judgement of many semi-skilled machine-operators like these, which was based on their accumulated experience on the job.[63] In the late nineteenth century, there was a category of workers in many industries, including iron work, known as "helpers," who were expected to have a similar level of competence as the new groups of semi-skilled; but, as their title suggests, they were primarily the craftsmen's assistants, whereas the semi-skilled were directly responsible for output at the centre of production. The dividing line between skilled and semi-skilled could also be much less distinct in mass-production jobs than the old distinction between craftsmen and helpers or labourers.[64]

The question might follow: at what point could "experience" be converted into the technical competence we call "skill"? This was hotly contested terrain in the early twentieth-century steel mill, between the owners who would never countenance the transformation of these occupations into anything like crafts or trades and the workers who wanted their indispensability recognized. In 1915 a board of inquiry into industrial unrest among the New Glasgow-Trenton steelworkers heard this confrontation in the testimony presented by Scotia's workers and managers. Several workmen voiced their bitterness at the low pay they received for applying what they called their "skill." One rolling-mill hand insisted that "skill and experience are necessary for rolls. It is a very important job. . . . All the work around the mill is skilled labor"; yet, he claimed, the average daily wage was a paltry $1.54. Repeatedly, the worker witnesses insisted that their jobs could not be done without some training. One asserted that "practically any branch of the work can't be done until you get into the way of doing it." Another related a story of a "green hand" working with him who lacerated himself on a hook after only half-an-hour's work. An axle-cutter thought an untrained worker trying to do his job would be "liable to do some damage," while a disc heater was certain an inexperienced worker "would not get production." In summing up the men's case, their spokesman, Clifford Dane, insisted that "the men on the mills are a class of expert and highly skilled mechanics." To hammer home his argument, Dane reported that the No. 1 mill could not work earlier in the week because three key rollers were absent: "With 2,300 men on the steel company's plant, these three men were so necessary that the mill would have to close down unless they went to work." Comparing their meagre wages with those of machinists and bricklayers, he declared: "These wages are not fair, are not just and moreover do not compare favorably with the wages of skilled mechanics whether in Nova Scotia or elsewhere in Canada." In their own estimation, then, these steelworkers were skilled men.[65]

The corporation's efforts to deflate these claims in the same hearings are revealing. Scotia's general manager intervened frequently to ask about how much explicit training these men got for their jobs. Most admitted that they had picked up their expertise informally by watching and learning from older workers. One explained: "I was often around there and saw how it was done. I was not like a man coming out of the woods and starting to work." The general manager hammered away at the lack of a formal apprenticeship: "you cannot compare a mill man with a machinist – one has served an apprenticeship, the other has not. You cannot put their earnings on the same basis for that reason."[66] In fact, most machine-operators in steel plants worked their way up job ladders into their more skilled positions. "The laborers are put into better positions as they are needed and as they show aptitude for the work," Scotia's president explained, and the personal histories of the men who testified before the board of inquiry, as well as of those interviewed years later in all the steelmaking centres, indicate that kind of progression through the ranks.[67] Steel companies could therefore claim, with some justification, that the self-styled "skilled" men in their plants owed their competence to their employer – after all, hadn't they taken farm boys and turned them into productive factory hands?[68] Their special competence, moreover, generally had no value outside the steel industry. Perhaps most important, they were vulnerable to replacement by ambitious men in the plant who had picked up the basics on the job through "being around there and seeing how it was done."[69]

Yet at the same time, the corporations knew what they would not admit to a public board of inquiry – that the knowledge of production still lay out in the plant, not in the front office. They had not been able to follow the dictum of F.W. Taylor, the prophet of "scientific" management, to gather up all that knowledge in management's hands and parcel it out to the workers. Scotia's president

sent his son to McGill University for professional training, but put
him on the end of a shovel in the firm's open-hearth department
and sent him off to broaden his experience in other plants in the
United States before fitting him into a managerial position at Trenton. When a young man began work as the Algoma president's
secretary in 1920, he found the corporation relying on a widely
used metallurgical manual put out by US Steel, but he also found
himself promptly sent out to spend a summer working amidst the
smoke and dirt to find out how steel was made. A few years later
another man arrived at Disco with a degree in metallurgical engi-
neering from McGill University, but was immediately put to work
as a labourer in the blast-furnace department, where he remained
for more than a year before assuming the position of superintendent. These shop-floor "apprenticeships" were normal in the steel
plants. In fact, many department superintendents, and some plant
managers, had worked their way up without any technical training
in a university.[70]

The additional evidence that steel management recognized the
importance of their semi-skilled workers was their effort to hold
onto them. Scotia was the first to build up a more settled work
force, hiring primarily local men from Pictou County and moving
many of them to Sydney Mines in 1904. As we will see in Chapter
Three, the other corporations also devoted increasing energy to
stabilizing an experienced work force. Stelco's vice-president admitted to a royal commission that "the longer you keep a man the
more valuable he is to you. I am talking from a purely selfish
standpoint, and believe if a man works for us and gets into the
habit of doing that work, see what it costs to replace him? Why,
there is everything to induce friendship to him." In slack periods,
the firms would go on short time to spread out the work, rather
than laying off many workers at one time and risk losing them.
Even after economic slumps forced the plants to shut down and the
workers to seek work elsewhere, the corporations would attempt to
recall the same men from distant points in the United States when
production picked up again. This practice was evident at Algoma
as early as 1904, when the general superintendent needed only five
days to reassemble a competent work force for the rail mill. "The
secret of this is that when the collapse came, Mr. Lewis, who is well
acquainted at various rail plants in the United States, went about
and secured employment for every one of his men, and right along
kept in touch with them," the *Sault Star* reported. "When Mr.

Lewis needed them again they were only too glad to come to him."[71]

Semi-skilled machine-operators in Canada's early twentieth-century steel plants, therefore, generally did more complex, responsible work than is often suggested by the term "machine-tending," and their employeers knew how important these men were to production. Yet their vulnerability and dependence on their employers for access to their jobs and the lack of portability of their self-styled "skills" deprived them of the independent power of craftsmen. Skill clearly means more than technical competence; it requires a social sanction won through the assertion of effective power. Craftsmen had struggled to obtain that grudging recognition by using their apprenticeship systems to control the labour market and their unions to police craft standards. In the process they had also drawn on the ideological resource of public respect for craftsmanship. Most steelworkers had jobs that lacked this kind of material and ideological leverage. Yet as the testimony at Nova Scotia in 1915 suggests, they were conscious of their value to their employers and bitter that it went unrecognized.

The mechanization of Canada's steel production process in the early twentieth century, therefore, had altered skill requirements in important, but often unexpected ways. Machines had replaced human muscle power in moving materials through the production process and had thus speeded up that process. But, on the whole, instead of breaking down old crafts, machines had simply replaced some older, heavier manual routines. Blast furnaces were much larger, but still involved fundamentally the same chemical process. Open-hearth furnaces had surpassed the limited output of puddling furnaces, but not the need for metallurgical knowledge. Rolling mills of massive proportions could handle much more metal, but still needed the careful attention of experienced workers. As a result, the old dichotomy of craftsmen and labourers had given way not to a mass of mindless machine-tending jobs, but to a new hierarchy of jobs with less gap between the least and the most skilled and with considerable demands for responsibility and judgement from workers filling them (hence, perhaps, with greater occupational homogeneity). Katherine Stone's emphasis on the stripping away of shop-floor discretion is plainly wrong. Close-knit, co-operative work teams at the centre of the production process headed by a few key workers like blast-furnace keepers, open-hearth melters, and head rollers exercised a range of independent decision-making on the job and a freedom from either direct supervision or machine-pacing that were similar to the shop-floor status of nineteenth-century craftsmen. For many more, the range of their discretion was much more narrowly circumscribed but nonetheless crucial to the most efficient operation of the new machinery. In fact, given the disappearance of so much unskilled labour, the *average* skill level in the new steel plants had probably gone up, not down.

And both the steelworkers and their bosses knew it. For the owners and managers of North American steel plants, the challenge was to keep these valuable men in their jobs in order to take advantage of their experience and thus maintain peak efficiency of production while keeping them in a state of dependency and power-lessness. For the steelworkers, the challenge was to find a way to turn their value to production to their own advantage. The technology itself had not determined the outcome of this struggle; it merely provided the new terrain on which it took place.

3

Labour Power

Implanting a brand new industry on Canadian soil posed several serious problems for the new steelmaking corporations. Once they had chosen a site, located adequate raw materials, arranged transportation facilities, and set up the appropriate technology, there still remained the problem of who would actually turn out the steel. At the turn of the century, the occupational category of "steelworker" scarcely existed in Canada. As in other mass-production industries, the steel companies had to assemble and shape a new work force capable of making their plants profitable enterprises. They had to draw from an unpredictable, often turbulent labour market and then had to impose some industrial discipline on workers with apparently unruly, recalcitrant, footloose habits. In such an unstable industry, it was essential for the owners and managers of the steel companies to turn their workers' labour power into the most productive labour possible. Their efforts resulted in an approach to industrial management that in some ways was specific to steel production but in many other ways was common to the emerging work world of early twentieth-century Canadian factories.

There were four identifiable phases in the development of managerial labour policies in steel plants in the period under discussion. Down to the severe depression of 1907-1909, the three newest steelmaking corporations were in a difficult start-up phase (Scotia faced the same problems in its new plant at Sydney Mines), but the federal and provincial bounties on primary steel products eased the growing pains. The end of that economic crisis marked the beginning of the second phase. With the bounties on their way out, all the corporations recognized the need for "reducing costs in order to keep the measure of profits up."[1] The great wave of technological innovation that hit the industry in the half decade before the war was accompanied by a new firming up of managerial policies. A third phase emerged toward the end of World War I in response to a phenomenal shift in the balance of power between bosses and workers, and involved a considerably more sophisticated approach to managing steelworkers. This unusual period lasted until the mid-1920s, when the crisis of labour militancy had passed. From that point until the late 1930s, the basics of pre-war management returned to the ascendancy, only this time with a more stable work force. To carry out these new policies, the corporations turned management into a more professional activity, modelled largely on American practices. But they also relied on some traditional, "unprofessional" techniques of controlling and disciplining a work force.

New Recruits

As we saw in Chapter Two, the technological transformation of the steel industry did not allow the new corporations to fill their plants with inexperienced "green hands." Like the nineteenth-century ironmasters, therefore, the steelmakers had to import their more skilled labour from countries with well-established steel industries. Scotia brought its first open-hearth melter from France and occasionally tapped the Scottish labour market, and Disco initially brought in some skilled German workers.[2] For the most part, however, Stelco, Disco, and Algoma looked to the United States for men to run the state-of-the-art American technology that was being installed in their plants and to work with the largely American managerial staff.[3]

Yet the Americans frequently would not stay in Canada,[4] and the steel companies soon discovered that they could not rely entirely on external recruitment for their skilled help. By the turn of the century, Scotia had already begun to look inside its own experienced work force for men to promote into the more important jobs (just as many American companies had done a few decades earlier),[5] and the newer corporations soon followed suit. Increasing numbers of the skilled and semi-skilled steelworkers, therefore, came from within the local, largely English-speaking population, although a trickle of recruits continued from south of the border. As a result, it was rare to hear the steel companies complaining of shortages of skilled labour (except when faced with the threat of a legislated eight-hour day, which, they argued, would necessitate an increase in their work force).[6]

Two groups of workers in an early twentieth-century Canadian steel plant, however, could not be recruited internally: the craftsmen needed in the mechanical departments and the unskilled labourers. For these, the corporations had to compete with other employers in external labour markets. Sometimes they attempted to organize their own internal apprenticeship programs for skilled tradesmen, but these were never adequate for their needs.[7] For the unskilled, they often tried to reshape the local labour market to their own requirements. Canada's steelmakers could have continued the nineteenth-century pattern of tapping local farm labour, and to a certain extent they did draw in Ontario and Nova Scotia farm boys, who were leaving home in such large numbers in this period.[8] But, since they soon found this source unreliable and expensive in the face of competition from other booming industries, they also began to look further afield. Canadian iron and steel producers who pondered the weakness of their industry in the 1890s recognized that their chief competitors south of the border had reduced their labour costs by hiring European immigrants and American blacks for the least skilled jobs in their plants. As a director of Nova Scotia Steel told a gathering of iron men in 1897: "If we in Nova Scotia were able to compete in wages with those in Alabama we could make iron as cheap as they do."[9] By the early 1900s each of the major corporations in the industry was trying to "compete in wages" by recruiting from new pools of cheaper labour.

Both Algoma and Disco were able to draw on the large numbers of Italian labourers brought in for local railway construction projects and for construction of the steel plants themselves. Francis Clergue had also sent directly to Italy for help in 1898 and to Chicago the next year. In 1902 Clergue's Consolidated Lake Superior Corporation established an Immigration and Labour Department to stock the local labour market, and 200 Italians who arrived that year were apparently part of that arrangement. It appears as well that one of Montreal's two leading *padrones*, Alberto Dini, was supplying Italian labour to the Sault.[10] Italians found their way into labouring jobs at the Sydney steel plant by a similar route. In this case, an enterprising contractor, Thomas Cozzolino, was the link in bringing in men from Boston, New York, and Montreal. At the same time, Disco's managerial staff was recruiting Hungarians and Alabama blacks from American steel mills. The corporation's blast-furnace superintendent arranged to pay the fares of several of these men and to set them up in company housing.[11]

Disco continued to find its isolated location a problem for maintaining a steady supply of cheap labour, especially in busy periods of full employment in the country generally. Periodically, therefore, the firm delegated labour agents to return to the homeland and round up workers. Both Ukrainian and Barbadian immigrants were brought to Sydney this way in the decade before World War I. The labour shortages of the war period stimulated a new search for cheap labour, and Disco arranged with federal authorities to get a good percentage of the 1,300 "Austrians" released from northern Ontario internment camps in 1916. The corporation's president even broached the subject of importing Chinese labour with the federal government.[12]

While better situated in the mainstream of the North American labour market, Hamilton's pool of labour willing to tolerate the long hours and unpleasant work in a steel plant could dry up in the boom years of the early twentieth century. As early as 1901 the *Labour Gazette*'s correspondent reported the Hamilton Steel and Iron Company's difficulties in obtaining sufficient unskilled labour "owing to the opportunities for employment for such men elsewhere and the heavy nature of the work in and about the smelter."[13] It was in this context of a tight turn-of-the-century labour market that the firm made an abrupt shift to employing Europeans by importing Italians from Buffalo, New York.[14]

Generally, however, these company initiatives were unusual after the start-up phase. Once the contacts had been made with particular ethnic groups, this transatlantic labour market operated through networks of intermediaries and labour agents who linked each ethnic enclave with potential employers. In 1914 the New Glasgow *Evening News* got hold of a circular published by the so-called "General Agency – The Transatlantic" in London, England, which the paper claimed was being distributed "by thousands in Russia." The flyer listed hundreds of jobs available in Cape Breton mines and steel mills, with good pay and lodging, and promised to arrange the appropriate documentation for immigration authorities. This sort of solicitation was not at all unusual. Moreover, as a particular employer developed an acknowledged preference for Europeans, informal networks of communication would carry the news back to the villages to direct future migrants to specific locations like the steel towns. The experience of one newcomer to Sydney was probably typical. He was a sixteen-year-old Pole living in a village in the Austrian Empire when in 1907 a letter arrived from his uncle in Sydney urging his mother to send the boy over. His mother borrowed $40 to buy the ticket, visa, and labour papers and sent the boy off in the company of three other men, with a supply of bread, onions, and garlic for the two-week trip to Quebec City. In that busy port, an interpreter made all the necessary arrangements for the train ride to Sydney. At 7 a.m. on the morning after he arrived in the steel town, the young Pole reported for work at Disco's plant.[15]

One important group of "foreigners" in the Cape Breton work force who reached the area largely through such informal contacts was Newfoundlanders, who worked in Wabana iron mines and Cape Breton coal mines, as well as in the steel plants at Sydney and Sydney Mines. By 1921 they made up 11 per cent of Sydney's population. In the eyes of the other townsfolk, they were often as strange and distinctive a group as the Europeans. "Their mode of speech was so different that it was difficult to understand some of them, depending on what part of Newfoundland they had come from," a retired Disco official later wrote. A railway contractor warned Scotia's general manager against hiring "these miserable half starved pirates," who he believed had proved to be "the meanest lot of devils and hardest men to get along with I ever saw"; Italian labourers, he warned, would be "Princes compared with New-foundlanders."[16]

Except at Scotia's Trenton works, these newcomers from conti-nental Europe and Newfoundland made up a large percentage of Canada's steelworkers in these early years. In 1910 a Nova Scotia commission of inquiry found Europeans and Newfoundlanders fill-ing half the jobs at Disco, while at Stelco the Europeans alone comprised three-fifths of the work force by the war. The percentages rose still higher during the wartime labour shortages. Some combi-nation of Italians, Poles, Ukrainians, and Hungarians seems to have predominated among the Europeans in all these plants.[17] These workers were not evenly distributed throughout the work force. The Europeans found work in less skilled labouring jobs in the yards and around the blast furnaces, coke ovens, and open-hearth furnaces. In Cape Breton the Newfoundland labour turned up in all these departments, especially around the blast furnaces, but also in the unskilled work in the rolling mills. In 1918 the superintendent of the Ontario Labour Department's Employment Bureau in Hamilton could report that "foreigners" did "practically the whole of the heavy and laborious work" in the city's iron, steel, and metal-working plants. As late as the 1930s this ethnic stratifi-cation of the work force was still intact. Europeans and blacks were virtually never allowed into skilled jobs, except perhaps in the steel companies' foundries. There was some clustering of ethnic groups

– Italians around the blast furnaces and blacks around the coke ovens, for example – but this appears to have been an informal process that never precluded other ethnic groups from working in these departments.[18]

What made these new ethnic groups so attractive to Canada's steel companies? In the first place, they had no long-term commitment to their jobs in North America and seemed willing to accept almost any terms of employment provided they could accumulate cash to send home. The historians of these southern and eastern European groups in Canada have recently been reminding us that, initially at least, most of them were migrants from distant villages where their peasant families required cash to cope with worsening underemployment, overpopulation, and agricultural depression. Overseas migration became an extension of patterns of seasonal migration within Europe intended to shore up a peasant way of life in serious crisis.[19] "One has only to be at the post office after pay day, and see men sending off hundreds of dollars to distant lands," a disapproving Sault Ste. Marie journalist reported in 1907. Five years later the same paper noted that $21,000 had been sent out from the Sault in money orders in one week.[20] Even those who planned to stay in North America seldom intended to remain wage-earners. In those cases, like many of the Ukrainians for whom immigration to Canada was more often a permanent project, the intention was to recreate the peasant life on a prairie homestead, and wage labour remained the same short-term goal of building up some cash savings. In most immigrant communities, moreover, small numbers of men (and a few women) settled in to provide the housing, provisioning, banking, or employment services required by their transient fellow countrymen. The same patterns were evident among the itinerant Newfoundland fishermen who arrived on a seasonal basis each year to take up labouring jobs in Cape Breton before returning to their boats in the outports.[21]

A stint in a steel plant might be only one stop along a vast continental trek in search of work. After conducting preliminary social surveys in Hamilton, Sydney, and four other Canadian cities for the Methodist and Presbyterian churches, Bryce H. Stewart noted in 1913: "Tired of 13 hours night shift in the steel plant at Sydney, the immigrant tries railroad construction out of Fort William, and when winter comes presses on to tend a coke oven or 'work a face' in one of the coal mines of Crow's Nest, or returns east to 'The Soo,' or Hamilton or Montreal or back to the Sydneys." A recent study of Italians in Algoma's early employment records revealed the general pattern: most stayed no more than five years and often worked for only a few months or even a few days at a time; of 252 Italians listed on the payroll in the last quarter of 1905, only twenty-five were still there by 1910. Before the war it was customary for many Italians and other Europeans to return home for the winter.[22] These patterns of working and living, in fact, strongly resemble what have come to be known as "guest workers" in twentieth-century Europe (and elsewhere).[23]

The so-called "foreign colonies" that appeared in Hamilton, Sydney, and Sault Ste. Marie strikingly reflected the transiency of these newcomers. Most of them were men in the the prime of life, and in 1911 women comprised only 19.5 per cent of the European-born population in Sydney, 22.8 per cent in the Sault, and 29.3 per cent in Hamilton. By 1931 the men in these communities still made up more than 60 per cent of the immigrant population.[24] On the streets immediately adjacent to the steel plants in each city could be found cramped boarding houses with tight rows of beds for the male labourers. In some cases, especially in the early years, the more isolated steel companies, Algoma and Disco, provided these boarding houses on land beside their plants, and, in fact, just after World War I, Algoma constructed new bunkhouses for several hundred workers needed for the anticipated expansion of production (which never happened).[25] But, more often, enterprising members of each ethnic group opened their doors to sojourners from their own homelands and filled every available space with beds. As a census taker discovered in Hamilton in 1911, the men would eat and sleep in shifts and "the beds never grow cold."[26] In 1913 Bryce Stewart's investigations turned up shockingly crowded housing among these boarders. In the shadow of Stelco's smokestacks, he counted 232 men, nineteen women, and twelve children living in seventeen eight-room houses; among these Italians, Bulgarians, Poles, Rumanians, and Macedonians, 213 of the men were single, either boarding or living co-operatively. In one block of Disco's company housing in the immigrant district of Sydney known as Whitney Pier, he discovered 331 Poles and Russians living in nineteen houses "jumbled into an area that scarcely exceeds an acre and a half," while in another block of nineteen private dwellings, 257 residents, predominantly Italians, shared the limited space. In both cases about half were male boarders. Many of Cape Breton's Newfoundlanders lived in similarly temporary accommodations, where their lifestyle in shacks on the edge of town, usually run by a few local shopkeepers, won them the label of "shackers."[27]

The European immigrants recruited to work in Canadian steel plants often lived in primitive housing, such as these shacks in Sydney's Whitney Pier in 1913. (United Church Archives)

These immigrant workers were enjoying some fresh air on the steps of their boarding house in Hamilton in 1913, where twenty-three of them shared thirteen beds in shifts. (United Church Archives)

Conditions in these boarding houses could often be as rugged as the bunkhouses in the logging and construction camps of the north and west.[28] One 1905 report in the Sydney press described the rough-hewn shacks that many migrant steelworkers inhabited in Whitney Pier as "bad beyond description":

> Some of the houses are fairly clean, but the majority are exceedingly filthy. There is no sewerage or water connections, and the ventilation is foul. Especially is this true in some shacks at the coloured quarters, which are occupied, not by negroes, but by Hungarians, Poles, and Newfoundlanders. The beds are simply a big deal table about seven feet broad, which runs down the whole length of the room. The men wrap themselves in their blankets and lie on this shelf as close as they can pack. There is also in the room a stove, which the inmates keep as hot as possible, and hanging from the ceiling, which is generally about ten feet high, are the spare clothes of the men.
>
> None of the beds are ever aired, for as soon as one shift is out of them, another is in. These are, of course, in only the worst of the houses.

The same year, Hamilton's Board of Health found similar conditions in the "foreigners'" shanties around the city's steel plant. Immigrants to the Sault in the same period later recalled equivalent housing. "We lived in shacks, a few boards nailed together, that's all," one man told a reporter. "They were all full of holes, the rain came in, the wind came in and the snow came in, and in the winter we froze."[29]

Such an economical lifestyle allowed these workers to save much of their earnings, but also to work for unusually low wages. A *Sault Star* reporter was struck by the dedication and thrift of the local Italian steelworker: "he lives on a few cents a day, and thus he saves sufficient out of his wages to make remittances home of $50, $75, $100 at a time. The average Irish or English laborer would on the contrary indulge in a 'blow out' or spree once in a while and be nothing at all ahead at the end of the year."[30] These migrant labourers, who were only expecting to work in a steel plant for a short time, were therefore cheap labour for the companies. "It would be a very serious matter to do away with foreign labour," a Stelco official warned in 1919 during heated debates about deporting the "aliens"; "if we expect returned soldiers to do the rough, rugged work, many of us would be out of business because we could not produce at anything like low enough cost."[31]

In the absence of payroll books for any of these corporations, it is difficult to know precisely how cheap this labour was, but it is possible to get an impression of unskilled wage rates from sporadic reports in the press and in government investigations. In the years between the turn of the century and World War I, the minimum wage rate of most Canadian steel-plant labourers hovered around 15 cents per hour, with surprisingly little variation during a period of soaring costs of living. This rate fell below the figure of 17½ cents set by Sydney's city council in 1910 as a "fair" minimum for labourers and was well below the 20 cents per hour that the federal Department of Labour estimated to be the average wage for factory labourers in Hamilton in 1915 or the 25 cents per hour earned by that city's civic labourers. The steel labourers' rate had generally more than doubled by the end of World War I, but it fell back by something like a third in the early 1920s. After the unskilled wage in the industry had edged up to an estimated 36 cents per hour by 1928, the rates were cut again at the beginning of the Great Depression.[32]

Europeans and Newfoundlanders continued to migrate to steel-plant jobs as long as work was readily available, but downturns in the economy could send them off in search of other jobs or back home. There was a great exodus after World War I,[33] and the combination of the interwar slumps, the winding down of alternative employment possibilities in railway construction or agriculture, and tighter immigration restrictions in the early 1920s and the 1930s undermined the patterns of sojourning in North America.[34] The percentage of "foreign-born" in the local population consequently began to decline. At the same time, however, growing numbers of these newcomers were choosing to stay in industrial jobs and put down roots in the steel towns. Gradually they found their way into better jobs in the steel plants, especially during the World War I boom and labour shortage.[35] In 1916, for example, Algoma reported to the federal government that it had 311 Italians working in semi-skilled jobs. Sixty per cent of the whole industry's foreign-born workers had been in the country at least ten years by 1931. The 1920s were a transition period in this process of settling down. Transiency had not disappeared: the 1931 census reported that 40 per cent of the steelmaking work force had arrived in the country since 1920 and that only about two-fifths of the European-born in the industry had become naturalized. The process was uneven across the industry, however. The Maritime steel towns saw a substantial shrinking of the European component in their steel plants, while Sault Ste. Marie experienced a more modest but nonetheless significant decline. In both cases, the frequent bouts of unemployment and regular underemployment in these crisis-ridden steel towns were no attraction to sojourners looking for quick cash. By 1926 an employees' brief to the federal tariff board could claim that two-thirds of Algoma's non-Anglo-Celtic workers were naturalized citizens of Canada. Hamilton, on the other hand, had a substantial increase in its European-born population during the 1920s. In general, the Ontario branch of the industry continued through the interwar period to have a higher proportion of workers from a European background than Nova Scotia. Transiency thus remained a dominant motif until economic disaster struck, first in Cape Breton and Sault Ste. Marie in the 1920s and then in Hamilton in the 1930s. By the outbreak of World War II, sojourning had long since died out, and Canadian steelworkers were preponderantly a home-grown breed, with proportionally far fewer Europeans in their ranks than had been present in the first two decades of this century.[36]

Besides ensuring a cheap labour supply, the transiency and inexperience of the new work force that had been recruited into the industry had had a second major advantage for the steel companies. Their lack of familiarity with steel mills and their vulnerability in a new cultural and industrial environment made them appear more docile and tractable than English-speaking steelworkers. One of Stelco's predecessors, the Hamilton Steel and Iron Company, must have had this concern in mind when it used European immigrant labour to help break a strike of its workers in 1902. That spring, about seventy Anglo-Canadian labourers around the corporation's blast furnace walked out on strike for a 10 per cent wage increase and promptly published articulate appeals in the press for public support. Rumours began to circulate that the men wanted to abolish the twelve-hour day and were taking the first steps toward unionization. It was no doubt this stubborn aggressiveness that would prompt a Stelco official to blurt out a decade later, "the English workman is the cause of more labour troubles than any other nationality." To combat such a spirit of resistance among its labourers, the corporation recruited strikebreakers, including a trainload of Italians from Buffalo, who were encouraged to camp in makeshift shacks on company property. The strike was broken, and the shanties became permanent bunkhouses for many of the firm's sojourning labourers. The shift to European migrant labourers in blast-furnace work was so thorough within five years that the next strike in that department in the spring of 1907 involved only "foreigners," mostly Italians.[37]

At the same time, since these workers had no long-term commitment to their jobs in Canadian steel plants and, before the 1920s, seldom stayed more than a few years at the most, their bosses could more easily intensify the labour process to extract the maximum from their labour power in a working day with a minimum of resistance. Not surprisingly, in 1910 a Nova Scotia commission found the heaviest concentrations of Europeans and Newfound-landers in departments on twelve-hour shifts, especially around the coke ovens and blast and open-hearth furnaces.[38] The mythology soon grew up that these new workers were necessary because Eng-lish-speaking workers avoided such jobs, but the nature of the work was in part shaped by the workers available. Probably only these sojourners could tolerate such exhausting toil. In fact, the work rhythms of peasant life had always involved seasonal bouts of intense, concentrated labour. Two American scholars noted in 1918 how Polish peasants in Europe had already adjusted these rhythms to earn as much as they could as quickly as possible:

> The peasant begins to search, not only for the best possible remuneration for a given amount of work, but for the opportu-nity to do as much work as possible. No efforts are spared, no sacrifice is too great, when the absolute amount of income can be increased. The peasant at this stage is therefore so eager to get piece-work . . . . They take the hardship and bad treatment into account, but accept them as an inevitable condition of higher income. When they come back [home], they take an absolute rest for two or three months and are not to be moved to do the slightest work. . . .

Similarly, a Scotia worker explained in 1919 how a transient worker would be "put on this piece work job and he knows that he is not going to be there probably for over two or three months, so he works very hard and spoils the job for the other men. . . ."[39] Provided there was a steady supply of this kind of labour, the steelmaking corporations would find it useful in their efforts to increase output.

There was one final bonus for the steel companies that hired these "birds of passage" – the cultural gulf that set them apart from the rest of the city's working class and inhibited class-conscious solidarity. These were men from peasant and outport backgrounds whose ties were usually stronger with family and village across the water than with fellow workers a few blocks away. Each of the "foreign colonies" had separate clusterings of fellow-countrymen, which generally kept some distance from each other, relied on their own networks of ethnic intermediaries, and occasionally scrapped openly on the job or in the streets. The presence of so many Europeans, moreover, deeply troubled the more skilled native-born and American workers, who charged that these "foreigners" had been "brought over to lower the standard of living."[40] There were a few incidents – surprisingly few, actually – of open conflict between these two distinct groups of workers before World War I, such as a 1912 strike precipitated by the introduction of two Poles into Stelco's wiredrawing department.[41] The war, however, ignited some vigorous nativist sentiments, partly based on the special attention given to all eastern Europeans by military authorities, but also often based on resentment of the better jobs the "foreigners" had acquired in the wartime prosperity and the higher earnings they were stashing away. By 1919 Anglo-Canadian workers in the major steel centres were agitating for exclusion of "aliens" from "white" jobs. "The Government should be told now and told plainly that restrictions should be placed on all cheap European labor," a Cape Breton labour paper argued that year; "especially should this apply to Germans, Austrians, Hungarians, and Russians." In Hamilton in February, 1919, a boisterous crowd of 10,000, in which "returned soldiers and working men seemed to predominate," demanded the deportation of "enemy aliens and other undesirables." Even more ominous was the appearance of the Ku Klux Klan in three Canadian steel centres in the 1920s, most notably in Sault Ste. Marie, where hooded bigots harassed the city's working-class European population between 1926 and 1929.[42]

As important as these anti-immigrant antagonisms were, we should bear in mind that there were no race riots comparable to those in British Columbia or the United States in the same period.[43] Ultimately, what mattered was not so much the open conflict between ethnic groups under one factory roof as their completely separate social worlds. An Italian blast-furnace worker, for example, would return to his boarding house, or perhaps later to his own home, in a continental European enclave where he enjoyed the support of his own community institutions distinct from those of the English-speaking workers. Each ethnic group retained its own language, developed its own mutual benefit societies (like the Santa Rita Society in Sydney, or the First Hungarian Workers' Sick Benefit Society in Hamilton), and patronized its own set of shops, cafés, community halls, bands, choirs, and theatres. Each worshipped in church congregations (mostly Catholic) of fellow-countrymen and celebrated its own ethnic and religious festivals. Each tended to send children to Catholic separate schools distinct from English Catholic schools.[44] What little Anglo-Canadian workers knew of life in these immigrant neighbourhoods came in large part from lurid press accounts of violence, drunkenness, and degraded living standards.[45] To most outsiders, the immigrant ghettoes remained exotic, alien, and threatening. Even as the "foreign" work force stabilized, their neighbourhoods continued to show the outward signs of severe poverty that English-speaking workers feared – overcrowding, poor sanitation, and general drabness.[46]

It is not only hindsight that has revealed how advantageous these walls of ethnic segmentation in the work force could be for the steelmaking corporations. In the wake of a large strike at the Hamilton Steel and Iron Company in 1910, a local newspaper noted how the transiency and strangeness of so many of the city's steelworkers left the public "comparatively indifferent to their claims." The "great manufacturing corporations" thus reaped a twofold benefit: "They get the work done at a cost less than the cost of getting it done by English-speaking workmen, and they prevent the enlistment of public opinion on the side of the workers when troubles arise with the foreign employees."[47] For these corporations, the interests of the migrant labourers in quick cash earnings meshed well with corporate strategies aimed at intensifying production, increasing productivity, and undermining potential worker resistance. But a common working-class community developed only with great difficulty in the circumstances of mutual suspicion and social fragmentation that resulted.

Modern theorists of segmentation in the working class have focused on the narrowing of the job hierarchy and concluded that homogenization was the main theme in the development of the labour force between 1870 and World War II, at which point segmentation began to emerge between core and peripheral sectors of the labour market.[48] This perspective misses the crucial divisions within mass-production industries along ethnic lines, which in Canada date from the turn of the century. Until the 1940s ethnic fragmentation was a persistent counterpoint to increasing occupational homogeneity.[49]

From the perspective of the migrant labourers themselves, a short-term job in a Canadian steel mill might be helpful for shoring up their peasant families' way of life, but the experience could be painful. Transiency could certainly lead to anonymity in the factory. In 1901 a Sault Ste. Marie shopkeeper was startled to find a man presenting a paycheque made out simply to "Italian, No. 151." A decade later, when asked about an illustrious Spanish anarchist who had allegedly worked at Stelco, Stelco officials admitted that "they could not tell whether such a man had ever worked there or not, as frequently foreigners' names are misspelled and more often they were known by numbers rather than by names." The same year no one could discover the name of an Italian who had been killed by a falling steel beam.[50] Even more threatening was the physical danger these peasant-labourers faced in a Canadian steel plant. "Several accidents have happened to foreign workmen who do not understand English at all or only very imperfectly," Nova Scotia's factory inspector noted in 1912, and five years later he linked their transiency with their accident rates: "Most of the accidents are due to inexperience, and happen mostly to unskilled labourers about the large plants."[51]

In the early twentieth century, then, Canada's corporate steelmakers drew together a new work force to turn out their steel. They found themselves compelled to rely on three different labour markets – two outside the firms (for tradesmen and labourers), and one inside (for most of the skilled and semi-skilled jobs in the mainstream of production). These workers were divided by more than their labour-market positions, however, since the steelmasters managed to change the ethnic complexion of the unskilled (and to some extent the semi-skilled) labour pool. These recruitment patterns followed the example of the American steel-plant managers, but there were distinctive Canadian variations. On the whole, it seems, although European migrants were present in large numbers, their impact was somewhat more limited, especially in Nova Scotia, as Canadian corporations tapped regional pools of underemployed rural workers, including Newfoundlanders. These workers were nonetheless as much "sojourners" as their European workmates until the industry's interwar crisis ended their casual participation in steelmaking. The work force was thus a complex mix with only limited stability or cohesion before the 1930s.[52]

Carrot and Stick

Assembling a suitable work force was only a first step in the management of steel labour. The corporations also had to find ways to make sure that they got the maximum efforts from their workers' labour power. What they needed, especially among the skilled and semi-skilled whom they entrusted with considerable responsibility for the rhythm and pace of production, was the means to make steelworkers work hard and fast. The new machinery in Canada's steel plants had reduced the corporations' reliance on human muscle power and thereby increased the volume, speed, and integration of production. But it had not given them "technical control" over the work force in the sense of rigidly pacing and disciplining workers.[53] Few steelworkers fit Charlie Chaplin's stereotype of the assembly-line worker in *Modern Times* struggling feverishly to keep up with the relentless speed of self-propelled machinery. Most of the individual machines in a steel plant, whether hoists, cranes, conveying vehicles, charging devices, or even furnaces, had to be activated by workers making their own judgements about timing and pacing. The steelmasters therefore had to find mechanisms for ensuring that these workers applied their labour power as intensively and conscientiously as possible. These managerial policies, rather than the machinery itself, promised to bring discipline to this new work force.

One technique for getting the most work out of workers was to keep them on the job for twelve hours a day (eleven on the day shift, thirteen at night) in the continuous-production departments – blast furnace, coke oven, open hearth, and, quite often, the larger rollings mills – and often among skilled maintenance workers as well. In 1910 an American study documented the clear policy of American steel-plant managers, beginning in the late 1880s, to lengthen their workers' hours on the job by systematically eliminating shifts of eight and ten hours and putting as many workers as possible on the long shifts. A 1919 study of the industry by the Interchurch World Movement revealed that the process continued in the following decade. These were the policies that American managers brought to Canadian mills. The twelve-hour day and seven-day work week (with a twenty-four-hour swing shift every two weeks) had the advantage of allowing the corporations to use a smaller work force than they might otherwise have required. Such a schedule also had a disciplinary power that Stelco's open-hearth superintendent explained to a parliamentary committee in 1910. In describing the long night shift, he noted that the men could do little more than work and sleep: "They prefer to work at night and then go home, go to bed right away and sleep all day. They get up at five or half-past five and go to the plant." This endless cycle of work made the men easier to manage. "We get better results from our men where we have them work 11 and 13 hours," the supervisor stressed repeatedly in his testimony; moreover, "the best men we have and from whom we get the best results are the men who stay at work at least 325, 330, or 340 days a year." Shortening the work week, especially by shutting down on Sundays and holidays, created "constant trouble" and "dissipation," he claimed. "It seems to give them too much time off; too much chance of spending money and to get around." Evidently, the long hours in the steel industry gave the corporations potential control over the whole lives of their workers.[54]

The steel companies repeatedly resisted efforts to introduce eight-hour-day legislation, arguing that they would have to search for many more skilled and semi-skilled workmen and increase their overall labour costs substantially. Usually they won reluctant support from the parliamentary bodies or investigative commissions examining the question. These long hours would thus last until 1930 in Hamilton and 1935 in Sault Ste. Marie and Nova Scotia – more than a decade longer than in American and European steel mills.[55]

Equally important for steel-plant managers was maintaining a vigorous pace of production during these long hours. In the nineteenth century, ironmasters had most often left direct supervision of the work process to their most skilled workers. Subcontracting the work to these men had been common in the British and American industries,[56] and in the early years at Londonderry and finishing plants like the Montreal Rolling Mills, but such practices seem to have largely disappeared by the time the Royal Commission on the Relations of Capital and Labour made its investigations in the late 1880s.[57] This was evidently a transitional point, in fact: when asked if he paid his own helpers, a Londonderry puddler explained, "It comes out of the puddlers' wage but the company pay them."[58] During these years, skilled workers nonetheless still had responsibility for directing the work of their less skilled workmates. Ironmasters indirectly disciplined these skilled work-leaders by tying their earnings to the company's economic fortunes: they were paid on a tonnage basis and on a sliding scale that reflected the selling price of iron products. This system was still used in a small department of the Hamilton Steel and Iron Company in 1910, but it was a quaint relic by that point.[59] Twenty years earlier, in the United States, the major steel companies had launched an all-out attack on this kind of co-operation with skilled workers' unions in the steel industry, culminating in a brutal crushing of worker resistance at Andrew Carnegie's plant in Homestead, Pennsylvania, in 1892. That historic defeat of craft unionism, combined with the ongoing technological innovations of the period, allowed the major American steel companies to break the connection between corporate earnings and skilled workers' pay. Henceforth the pattern – soon copied in Canada – was to establish wage scales from the base rate for unskilled labour.[60]

Along with the decentralized management of late nineteenth-century iron manufacturing had gone considerable sloppiness and informality about monitoring production costs. At Londonderry in the 1870s, the manager had maintained a constant battle with the furnace-keeper to record his daily costs in a central ledger, and in the Ontario Rolling Mills Company, cost-keeping was similarly casual in the 1880s, according to a company history: "While due regard was paid to the price of materials purchased and to the money volume of sales, little actual record was kept of the costs of operations, and what clerical staff there was concerned itself rather with correspondence and other immediate affairs than with the control of the use of working capital."[61] After 1900 the new Canadian steelmaking corporations were too obsessed with cutting costs to tolerate this kind of hit-and-miss bookkeeping. Like their American counterparts, they all developed a program of meticulous cost-accounting. "Every ton of this immense turnout is manufactured on a solidly economic basis," Algoma's president told the press in 1905. "We have the cost of production figured to the highest degree of satisfaction." Procedures for monitoring costs became centralized, and highly bureaucratized reporting routines kept all members of the managerial hierarchy on their toes. The most visible symbol of the new bureaucratic age of work relations was undoubtedly the time-clock, first introduced in Canadian factories in the early 1900s, which brought complaints from Disco workers soon after the plant opened.[62]

Yet, while all the Canadian steel companies used a bureaucratic framework for establishing clear lines of managerial responsibility and accountability, this movement toward "systematic" management so characteristic of early twentieth-century industry did not involve rigorous intervention into the day-to-day operations of individual departments. In discussing the American steel industry, Katherine Stone credits Frederick W. Taylor with a reorganization of managerial practice to break down skill, remove from workers the conceptual function in the production process, and place it in the hands of front-office planners, with his celebrated time studies, rate-setting, and incentive wages.[63] Subsequent research by Daniel Nelson has raised serious questions about the impact Taylor could have had on steel-plant management, since his experiments were confined to the machine shops and soon-to-be mechanized labouring work and never touched the main arenas of iron and steel production.[64] Similarly, no evidence has survived of white-collar men with stop-watches roaming about amidst the smoke and din of Canadian steel mills. Top-level managers arranged the overall layout of production processes, established general manpower requirements and employment policies, and closely monitored the level and the cost of the output. But for setting the pace and rhythm of daily production, they relied on older patterns of less centralized management by front-line supervisors. Taylor's theories of stripping foremen and lead hands of all conceptual responsibility generally met a deaf ear, probably, once again, because iron and steel production continued to require workplace know-how that white-collar managers with professional training or other business experience could never fully master.

Managerial methods in early twentieth-century steel plants, therefore, incorporated versions of two familiar approaches to stimulating workers' output – the carrot and the stick. The old sliding scale of wages had once been an effective incentive that relied on the spur to an ironworker's self-interest. In the new steel plants, managers cut the link to the size of corporate profits but turned to new incentive schemes that they hoped would have a similar appeal to workers. In the start-up years at Algoma and Disco, many workers seem to have been paid simply by the day. But in an effort to cope with the severe depression of 1907-1909 and the impending demise of the federal bounties on steel, both firms turned aggressively to new systems of wage payment based on tonnage, as Stelco and Scotia had been doing for years. In the 1920s Algoma explained to a government commission how the steel companies had introduced not only tonnage payments but also the new scheme of "premium bonuses" on tonnage beyond a specified minimum to stimulate workers' effort without paying them straight piece rates: "it is the general practice to give every man directly concerned in production a definite interest in the tonnage produced." Workers could be expected to discipline each other to keep up a fast pace, since most tonnage rates were paid to production teams. By 1923 Stelco's blast-furnace superintendent could describe "good incentive wages" as one of the keystones of the firm's industrial relations policies.[65]

Superintendents and foremen were also under steady pressure to increase the daily output beyond the rated capacity of the steelmaking facilities. In April, 1906, for example, Algoma proudly announced that its rail mill had set a new record: 859 tons of rails had been turned out in twenty-four hours, on equipment built to produce 500 tons per day, and Disco's previous record had been smashed by fifty tons. By September the daily output had passed 1,000 tons and seven years later reached 1,400. In 1930 Algoma boasted a new high of 1,670 tons. Both plants continued to report "record months" in all departments.[66] For foremen and superintendents, there was often a bonus plan designed to reward big increases in output in their departments. The competitive element in this process also played on the steelworkers' pride and sportsmanlike machismo. As one old-timer explained: "you tried to push and when they say let's break a record today and everybody would be out for that record. Oh, the odd one might hang back, but most people would say let's go, and when that whistle blew you tried to give another ingot to them to try to break that record." To honour the occasion, supervisors would usually hand out cigars, but, as the broken record soon became the new norm, the exhilaration of accomplishment was no doubt soon forgotten in the new strain to keep up.[67]

Table 2
Average Wages in Selected Canadian Manufacturing Industries, 1917–1950

	All manufacturing	Agricultural Implements	Autos	Electrical Parts	Printing & Publishing	Pulp & Paper	Steel
1919	$938	$1,047	$1,397	$884	$989	$1,230	$1,246
1925	971	1,101	1,577	992	1,305	1,267	1,325
1930	1,001	1,145	1,422	999	1,459	1,221	1,449
1935	874	962	1,321	906	1,275	1,143	1,247
1940	1,084	1,199	1,781	1,123	1,397	1,475	1,566

SOURCE: *Canada Year Book*, 1919-40.

Of course, incentive wages had less appeal if production was too uneven. The frequent layoffs and shutdowns in the Canadian industry certainly hampered the steelworkers' ability to make the most of the incentives. And investigative commissions heard complaints from workers on piecework, especially in the rolling-mill departments, that the pattern of small-batch production for the limited Canadian market, requiring frequent changeovers and set-ups, undercut their earnings.[68] Yet for workers who held onto their jobs and who were not plagued by the disruptions of short production runs, the long hours on piecework could pay off. Federal government statistics, summarized in Table 2, indicate that steelworkers' average annual earnings were always well above the average manufacturing wage and even above the earnings of the largely skilled workers in printing. Among mass-production workers, only those in the automobile industry did better (like steel, this industry had few women workers whose wages would drag down the average figure). A Scotia employee and union leader had to admit in 1919 that his fellow workers had accepted piecework "as an opportunity to earn higher wages even if they have to labour in a slavish manner. . . . Men here earn at times as high as nine or ten dollars a day but they do as much work in a day as an ordinary man might be expected to do in three. . . ."[69] Of course, the external stimulus was not simply avarice but the chronic insecurity of income and employment. In an age before social security legislation, joblessness, accidents, illness, or death could throw a working-class family into crisis. It was necessary to make hay while the sun shone.

The "carrot" of incentive wages could never have been enough alone to keep the steelworkers working, however. After all, the men might have set their maximum output, just as nineteenth-century craftsmen had frequently done.[70] In fact, as we will see in Chapter Four, during World War I, workers did not maintain the productivity expected of them once full employment in the wartime economy removed the whip of threatened poverty. Steel-plant owners and managers had to look to sterner measures for maintaining and increasing productivity. For the most part, they placed the "stick" of discipline in the hands of front-line supervisors. The superintendents and foremen were given responsibility for pushing workers to reach their peak productivity and for maintaining discipline and respect for managerial authority. Their role was not so different from that of many nineteenth-century foremen, but the pressures from cost-conscious senior managers undoubtedly intensified their anxieties and, in turn, brought the screws down still harder on the workers in their departments. A harsh, blunt, often abusive authoritarianism resulted from this so-called "drive system" and, with the partial exception of the small groups of more skilled workmen in some areas of the plants, these departmental despots ruled their industrial bailiwicks with iron fists.[71]

In part, their authority was a matter of theatre. Most of these supervisors developed reputations for gruffness, profanity, and physical prowess, which was intended to instil fear and deference, and which was laced with a strong dose of racism toward the non-English-speaking workers. Yet they also had real, tangible power to wield over the workers in their departments. Despite the top-level administrative reforms of the new systematic management, superintendents and foremen were left with a surprisingly wide range of authority. Initially they had clear powers to hire, promote, and fire workers,[72] and even after the creation of more centralized employment offices in each of the plants just before the war, they held onto that authority in practice through quiet instructions to the personnel departments. Virtually all the oral history covering the work experience in Canadian steel plants before World War II, from both management and labour sources, has revealed that front-line supervisors had not sacrificed one bit of their effective power in this area.[73]

Favouritism flourished in such circumstances. Most commonly, English-speaking workers needed some connection or sponsorship inside the plant. Soon after Disco's plant opened, workers were complaining loudly about the preference shown to Americans in hiring, promotion, and layoffs by the American supervisors that the corporation had brought up to oversee production. Those patterns persisted. One steelworker at Sydney remembered that his brother, an electrical worker at Disco, had enough pull to get him hired in 1922. "I think you had to have influence somewhere to get on as an apprentice," he reflected. Another worker from Algoma described how a neighbour who was a department foreman arranged to get him hired in 1928, even though 200-300 unemployed men were milling about in front of the gate by the employment office every day: "if you were a good friend of the foreman or a good friend of the superintendent and he sent word to the employment man at the gate 'Send me such and such a boy,' in you went, and otherwise you could have went there day in and day out and never got a job." Another retired Algoma worker explained that for "almost any of the steady jobs in the plant there was a lot of family preferences." The same was true at Stelco, according to a man who got his job there in 1936 because his father was the foreman: "The Anglo-Saxons were in there as brothers or cousins. I was kept on, even though others with years of experience were being laid off, only because I was Bill Martin's son."

Fraternal lodges could be equally crucial for getting hired and getting ahead in a Canadian steel plant. In Sydney, one retired steelworker explained, the mechanical department was dominated by the Masons, the foundry by the Oddfellows, and the open-hearth by the Knights of Columbus. "Your chance of promotion depended to a very great extent on your association," he recalled. "I know in the machine shop it would be embarrassing sometimes to see the people that would be promoted simply because of the ring or pin they wore and a better man would be laid aside." Another Sydney worker made similar use of his connection to the church that his supervisor attended.[74]

Europeans, blacks, and even migrant Newfoundlanders most often had to pay for such opportunities, even for the most unpleasant jobs on the bottom rungs of the occupational ladder where they started. A 1910 strike at the Hamilton steel complex first brought to light the practices of foremen in demanding a fee for the job from immigrant workmen and a weekly retainer for holding onto it. The corporation made a great show of publicly decrying such practices and firing two foremen involved in them, but by the 1920s and 1930s these practices were nonetheless flourishing in all the steel centres. In 1921 unemployed steelworkers in Sault Ste. Marie publicly denounced foremen who "were in the habit of taking booze, cigars, and money for giving employment," and a Hamilton unionist reported "a great deal of talk floating around the mill about some of the members being able to buy soft jobs." The memories of retired steelworkers are full of stories about these demands for money or goods and services. Foremen routinely insisted on bottles of liquor in return for guaranteed employment or re-employment after a layoff. At Disco "the foreign people had to buy the job," said one retired Ukrainian steelworker who started there in 1933, but that was not the end of the tribute to be paid to supervisors: "Well, if you wanted to work steady in Sydney, you had to give the boss a bottle every week. . . . There even was an agent in the open hearth and the mixer, he collected two dollars from each one there and gave it to the boss, so those fellows could stay on the job." A Stelco employee remembered a rotation system where a different worker brought the foreman a bottle every day. On other occasions a worker in the plant would be expected to "stick five dollars in a pack of cigarettes and hand it to the foreman." At Algoma, the bricklaying department (in charge of relining furnaces) was "the worst of them all," according to one old-timer who started there as a bricklayer's helper in the 1920s and had to keep the liquor flowing to his foreman if he wanted a promotion to bricklayer. That arrangement for promotion would work "until you fell out with him and then you were finished." An Italian blast-furnace worker in the same plant described to an interviewer how in the late 1920s he and his best friend had carried a big bottle of whisky, a goose, and a duck to a foreman to get his friend a job. Clearly these practices ran through all the departments where immigrant workers were found.[75]

Besides retaining the power to hire, promote, and fire, superintendents and foremen were allowed a remarkable flexibility in setting wage rates for particular jobs. In 1903 Disco admitted that foremen had this power, while by 1915 Scotia explained that its foremen still determined workers' individual wage rates, except "when important changes are made, in epoch-making crises, which only occurs every few years. . . ." By the end of World War I, steelworkers were complaining about the multiplicity of rates paid for the same work in a department. A Sydney worker later recalled that "if you were a good fellow with the boss, well, he'd give you a few more cents an hour. It wasn't a man's ability at all – just being a good fellow."[76]

Amidst such arbitrariness and uncertainty, workers could live in dread of losing their jobs. One Algoma worker who had bought his job was afraid to report his hernia to his foreman lest he not be rehired after recuperating. Supervisors used this anxiety to get themselves invited to weddings and other celebrations. Stories have survived about the two men who ran the Algoma employment office, who "used to go around the west end on visits, and . . . come back half tight or loaded with cheese and salami and everything else." Occasionally this tyranny could extend to demands for the sexual favours of workers' wives. All workers felt the insecurity of this system of management. In the words of a Sydney steelworker, "the boss would send you home if he didn't like the colour of your hair, if he didn't like the church you went to, if he didn't like the way you voted on election day. He could send you home and there was no questions asked." Another man concluded with deep bitterness, "The foremen were kings!"[77]

By the end of World War I the steelmaking corporations were aware that, in the words of a Stelco vice-president, they had "more trouble through workmen and foremen than anywhere else," and began to explore ways "to educate a better class, a higher class of foremen . . . as to what was fair between employer and employee and to produce the best results." Algoma launched the most elaborate program of foreman training (or retraining). "We have spent a year or more in weeding out incompetent foremen and checking up others who were indifferent, and educating other men for the position of foremen," the general manager explained in 1921. But these efforts do not seem to have borne much fruit or to have lasted much beyond the end of steelworker militancy in the early 1920s. Consequently, the same kind of informal but effective power that had irked new management theorists like F.W. Taylor hung on until challenged by the workers themselves through new industrial unions in the 1930s and 1940s.[78]

The corporations also developed more general policies of repression that applied across their plants. Discussing or attempting to organize a union could lead to instant dismissal, and the aftermath of unsuccessful strikes usually involved the blacklisting of ringleaders.[79] By World War I, moreover, the corporations were using a network of secret spies to keep close tabs on any sources of trouble and to report to the regular company police. The notorious American industrial-espionage and strikebreaking organization, the Thiel Detective Agency, was at work for the corporations in Sydney, Hamilton, and Sault Ste. Marie by the end of the war. In times of labour militancy, as in 1919 and 1923, the corporate security systems would also collaborate closely with the state's police and military. The provincial police records of both Ontario and Nova Scotia contain copies of these reports.[80] All of this repressive apparatus in Canada's steel plants was intended to prevent the development of any collective power among workers that could effectively challenge managerial authority. The entire administration of the corporations' labour policies was premised on the steelworkers' complete subservience to their bosses.

The discipline exercised over Canada's first steelworkers, then, was much less the product of new managerial theorists than the intensification of older methods of prodding and goading workers to produce quickly and obediently. Appeals to workers' individual self-interest in the form of incentive wages were buttressed with ruthless administration by front-line supervisors, who, whether or not they were always tyrannical, held their workers in bonds of personal loyalty (and, most often, fear). Fundamentally, these practices rested on manipulation of the fear of unemployment and poverty that a large pool of surplus labour outside the plant gates made possible at most times. Despite these continuities with the late nineteenth century, however, there were new elements in steel-plant administration. While pure-and-simple scientific management had little impact on shop-floor administration, the trend toward "systematic" management inside the large new corporations, especially the development of cost-accounting, put much more pressure on the lower-echelon supervisors and thus on their workers. And, of course, the oligopolistic power of Disco, Scotia, Algoma, and Stelco gave them much greater control over the steelmaking work force that depended on them for jobs, so that wielding repressive techniques like blacklisting could be much more effective in curbing collective action among steelworkers. Certainly, the combination of disciplinary structures in the steel plants would encourage a worker to keep his mind on production and to choke down any disgust or discontent that working in steel might raise.

The Web of Dependency

The dirt, danger, long hours, and harsh supervision made early twentieth-century steel plants anything but attractive workplaces. Workers might well be expected to push off in search of other, less demanding jobs. The steel companies had few worries about the departure of their unskilled labourers, as long as there was a steady supply of them flocking to the factory gates. They could not treat the services of the more skilled men so cavalierly, however. As we saw in Chapter Two, the new production processes relied on a seasoned staff of permanently employed, experienced, and responsible steelworkers. In addition to the problems of recruitment and discipline, therefore, the new corporations had to work at building morale, loyalty, and commitment among their most valuable workers. Policies had to be developed that tied them to their employers without at the same time conceding their importance, in order to avoid the re-emergence of the kinds of independent craftsmanship that had characterized nineteenth-century iron work. The key seemed to be ensuring their dependence on the company.

The companies' policy of internal recruitment for the better jobs was an important mechanism for drawing steelworkers into this web. One of the only breakdowns of steel wage rates still available (in Disco's blast-furnace department in 1919) indicates that a large number of job rates were bunched together at a relatively low level (mostly between 30 and 35 cents an hour).[81] Competition might be keen for the jobs that paid a few pennies more per hour and held out the prospect of more regular employment. As one of Nova Scotia's radical steel unionists would argue in 1922, this experience set them apart from the less hierarchical world of the coal miner: "One worker on the steel plant was pitted against his fellow worker in the struggle for advancement, and [in the context of their employer's hostility to unions] this advancement into a better job was the only relief the steel worker had in sight."[82] It should not be assumed, as several recent writers have claimed, that these "job ladders" were created specifically to control and discipline labour in the plants.[83] More careful research has revealed that it was the less skilled workers on both sides of the Atlantic whose demands for access to the more skilled jobs turned internal recruitment into a regular practice.[84] Employers co-operated because they had discovered that recruiting from within allowed them to draw on a pool of workers who had passed through the informal "apprenticeship" of watching and helping the more skilled men and who would therefore need little additional training. The disciplinary power of this system of promotion only became apparent over time, as supervisors exercised their discretionary power. The internal job ladders over which the foremen and superintendents presided certainly encouraged men who hoped for advancement to stay in their jobs, to curry favour, and to avoid any challenges to managerial authority (like discussing strikes or unions).

From soon after the turn of the century, the corporations also undertook to meet certain of the workers' basic social welfare needs. While job ladders may not have been designed with a purely manipulative intent, the same cannot be said for the increasingly sophisticated package of corporate welfare that the steel companies began to make available. There was an element of public relations in this activity, but before World War I the Canadian steel companies escaped the kind of intensive public scrutiny that hit their American counterparts in the form of massive social surveys and congressional investigations.[85] The Canadian companies were unabashed in admitting that their primary goal was the contented, settled worker. Stelco's vice-president bluntly told the Royal Commission on Industrial Relations in 1919 that welfare programs were not acts of philanthropy toward the firm's employees but were intended "to give them a direct interest in the business and promote continuity of employment," since "continuous and contented service is an asset to any company."[86] As we will see, worker discontent could take two disruptive forms – informal resistance through work slow-downs, absenteeism, or labour turnover, and collective action through strikes and unionization. Corporate welfare was aimed at both these forms of working-class insubordination. Most often, it was also beamed at the more skilled, English-speaking segment of the steelmaking work force, whose knowledge and experience made them so valuable to the companies and so ripe for organizing. Indeed, most of these programs were announced in precisely those periods when steelworkers' discontent was boiling over.

In some ways this new so-called "welfare capitalism" was not so different from the paternalistic policies of nineteenth-century iron-masters, who built small communities around their blast furnaces and forges and supplied their workers with housing, stores, even churches and schools.[87] These older practices had a similar intent of stabilizing a scarce work force.[88] Yet this old paternalism had usually radiated from a single entrepreneur, an ever-present community leader who ruled his works and his workers directly from some large baronial home overlooking the industrial village. Nova Scotia Steel first appeared in the late nineteenth century as this kind of employer, with the Fraser family providing the entrepreneurial leadership.[89] In the early twentieth century, large corporations with their more impersonal boards of absentee directors turned to welfare capitalism in the hopes of creating similar bonds between employer and employees without the personal touch of the owner-manager.

The first, most traditional, and, in many ways, most basic form of corporate welfare was company housing. Initially, like job ladders, providing accommodation for workers had a practical dimension dictated by the local labour market. In Sydney, Sydney Mines, and Sault Ste. Marie, there were too few private residences for the thousands of new steelworkers.[90] To hold onto their work force, therefore, Algoma and Disco threw up bunkhouses for the most transient. For managerial and clerical staff and for more skilled workers, they also arranged for the construction of better-quality, family-sized houses to be rented out, just as they were making available to the corporations' miners in adjacent communities. In some cases, workers were also encouraged to buy these dwellings on instalment plans. Scotia was particularly committed to this method of stabilizing its work force.[91]

This kind of company housing then existed to meet an immediate crisis of accommodation and to encourage steelworkers to settle in the steel towns. But it, too, came to have disciplinary power for management, since workers could find themselves homeless for insubordination. On several occasions, Disco evicted strikers from the corporation's houses. A Sydney steelworker described to the Royal Commission on Industrial Relations in 1919 how a man who "became distasteful or unsatisfactory" to the company would have to dispose of his house and leave town; "and when a man has spent every cent he has earned as a laborer for fifteen years to pay for his home, he cannot always get out so easily."[92] The impact of company housing on Canadian steel towns should not be exaggerated – Stelco had no need for it, and in neither Cape Breton nor Sault Ste. Marie did the corporations' houses completely dominate the local market for rental accommodation – but it did play a role in drawing considerable numbers of steelworkers into a state of dependency on their employers.

The corporations had other mechanisms for encouraging their workers to identify their long-term economic security with their employers. Well before World War I, each of the steel companies sponsored a mutual benefit society to provide insurance against sickness, injury, and death, on the model of the insurance plans of fraternal societies and (more importantly) trade unions. Typically, the corporations provided a quarter to a half of the funds and controlled distribution of the benefits.[93] To obtain their benefits, Disco employees had to appear at the office of their society's secretary, who also handled the corporation's rental housing.[94] The relief society's administrator could likewise have responsibility for arranging loans to employees, covering anything from mortgage or medical payments to groceries. In an age before social security legislation, the corporations' welfare might be, in the words of a retired Algoma steelworker, "the only type of security you could have."[95]

Before the war the only other welfare services provided by most of the steel companies were basic first-aid and medical facilities to deal with the rising incidence of accidents.[96] Stelco took the only step beyond this limited program by introducing a modest profit-sharing plan in 1913, modelled on a similar scheme run by us Steel. Workers earning less than $600 a year (that is, the vast majority) could buy only one share, but "as an inducement to each employee to make his [monthly] payments regularly and to remain continuously in the employ of the Company," the corporation promised to declare a bonus on each share, amounting to the princely sum of five dollars. Six years later the company's vice-president estimated that some 629 employees, out of perhaps 5,000 on the payroll, had bought into the plan. "Continuous, faithful and willing service is among the principle assets of any industrial firm, and this is certain to be stimulated by a sense of partnership," opined the industry's trade journal on assessing Stelco's plan in 1923.[97]

Until the end of World War I, Canada's steel companies had a relatively limited set of welfare programs compared to their counterparts in the United States. But, by that point, their workers' unprecedented challenges to managerial authority and a more menacing public opinion were forcing owners and managers into some serious soul-searching about their managerial policies. Like so many other Canadian employers in this period, the steelmaking corporations correctly sensed that they were locked in a battle for the allegiance of their workers and for the legitimacy of their institutions of private capital accumulation more generally. In their plants they had to cope with labour turnover of staggering proportions, absenteeism, reduced "effort" on the job, strikes, and surging industrial unionism. Outside they faced widespread criticism of corporate insensitivity in Canadian society generally – from journalists, academics, increasingly powerful farmers' organizations, a buoyant new labour movement, even from the highest councils of the Methodist Church, which passed a resolution in 1918 calling for production for service rather than for profit.[98] In this context, Canada's steelmaking corporations, like their counterparts in many other industries, began to fashion a new notion of management that allegedly took account of the "human factor" in production. Distinct offices or departments of "Industrial Relations" appeared for the first time, and each company introduced well–co-ordinated corporate welfare packages that widened the range of economic security and sought to build loyalty and commitment to the firm among steelworkers. What made these managerial initiatives such a substantially new departure was the explicit connection made between contented, loyal workers and productivity.

In part, the new programs were simply an expansion of pre-war provisions for workers' economic welfare. In 1920 Stelco announced a new pension plan, and Disco followed suit early in 1923.[99] The lengthy service requirements for eligibility – twenty-five years at Disco (which had been producing steel for only twenty-two years!) – were once again intended to coax employees into settling down in the firm. "Industrial pensions are found to increase the efficiency of labour by promoting contentment, loyalty, and 'morale,' " Besco's superintendent of industrial relations wrote in 1928; they were also, he added, "more economical than the practice of retaining old employees on full wages after they have passed their maximum efficiency."[100]

The new welfarism of the post-war period, however, reached out to steelworkers with appeals to more than their own self-interest. Each of the steelmaking corporations undertook concerted efforts to boost morale and promote a sense of belonging to a corporate team or family that would encourage workers to give up their footloose habits and to ignore the alternative spirit of working-class solidarity in unions. In 1920 Stelco and Algoma both began issuing monthly magazines, *Stelco News* and the *Algoman* respectively, which contained news from around the plants, reports on safety work, social and personal notes, and technical articles. "The aim of this publication," the editor of *Stelco News* explained, "should be to represent every worker in the plants, and make them realize that THEY constitute the Company – that their interests are also the Company's interests." After the tumult of 1923-1925 in Cape Breton's coal and steel towns, the British Empire Steel Corporation launched a weekly publication, the *Besco Bulletin*, which the firm hoped would help to make it "an institution to be proud of, an institution to be pointed to as an example of good fellowship and loyalty and united endeavour." The federal Department of Labour praised such publications as "one of the most effective agencies for the restoration of that direct contact between the employer and his work people which formerly existed but which has largely disappeared in the vast and complicated industrial enterprises of the day."[101]

Company recreation programs were similarly expanded. Corporate teams had been playing on the fields and rinks of the steel towns since well before the war, but these were now reinvigorated, and departmental competitions blossomed in every sport from bowling to hockey. Algoma went so far as to rent Sault Ste. Marie's YMCA building and turn it into a "Steel Plant Club" with recreation and sports programs, a cafeteria, and numerous night classes taught by company officials. These activities got full, appreciative coverage in regular columns of the local *Sault Star*.[102] Disco similarly announced the creation of an "employees' service division" within its new Department of Industrial Relations, which would oversee the various welfare plans and much more: "The plans for this division contemplate district nurses, hospitals, garden plots, Company's farm, employees' clubs, athletic, musical and dramatic societies, employees magazines, boarding camps, cafeterias, restaurants, and the accommodation for single-men boarders." Most of these grandiose plans, however, never saw the light of day in Cape Breton.[103]

Nothing, however, had quite the symbolic importance of the new safety campaigns that each of the major steel companies launched right after the war. Modelling their work on an elaborate scheme that US Steel had developed before the war,[104] Algoma, Stelco, and Disco announced "Safety First" programs for their plants that went far beyond the band-aid medical services they had previously made available. The programs were essentially the same in all the steel plants, but Algoma's got the most publicity. In 1920 the corporation's new general manager, a former US Steel official from Gary, Indiana, hired an enthusiastic newspaperman, Frank J. McGue, as "Superintendent of Industrial Services and Director of Safety" to head up the new safety work. McGue organized an extensive publicity campaign and arranged for new inspection and reporting procedures from supervisors. "Appeals were made to the pride and interest of the superintendents, foremen and the men by means of publicity in the local press, in the publication of the Corporation, by posters in the way of mottoes and pictures, translated into the languages spoken by the workmen, and tacked up wherever available, and by safety instruction," the *Labour Gazette* reported. Competitions were set up between departments to encourage "accident-free" months. Eventually safety committees, appointed by supervisors from workers on the shop floor, were established to monitor safety on a regular basis. Within the first year reported accidents had been cut by 60-70 per cent. All the Canadian steel companies continued to report significant reductions in accidents during the 1920s.[105]

The advantages of such a program for the embattled corporations were apparent from the start. Certainly the reduction of accidents made good business sense – workers' compensation costs would be lower and the absenteeism caused by workplace injuries would be decreased. The great emphasis placed on workers' "sense of individual responsibility" and careful concentration on their work to avoid accidents held out a similar promise of increased productivity, as did the network of departmental watchdogs on the safety committees, who could be expected to exercise some discipline over the work habits of their fellow workers. But it was the ideological power of the safety campaigns that made them so appealing. Here was the opportunity for large corporate employers to show a humanitarian concern for their workers, in order to win back their confidence and allegiance and to counterpose to labourist or socialist critiques an alternative vision of social welfare within corporate capitalism. The power Algoma's president gave to McGue and his inspectors to investigate and publicize unsafe practices created enough squeals of outrage from front-line supervisors that workers might have been convinced that a new age of enlightened management might be dawning. The corporation's general manager liked to talk in grand terms about the "new doctrine of safety" and "the big Safety Movement that is sweeping over the world."

Propagandists for "Safety First" claimed that it could lay the basis for a new accord between workers and bosses. In 1920 the Canadian Manufacturers' Association journal reported favourably on a speech delivered to its Ontario membership by an American safety expert. He had predicted that "the men who were working together in the great humanitarian cause of safety would extend their field of vision to include other matters vital to the welfare of the industry and on which differences of opinion might exist," and that "the safety council would, indeed, prove to be a real safety valve when trouble threatened in a plant." Algoma's Frank McGue similarly believed that safety was "the foundation stone of all industrial relations, of all true, honest and sincere concord and understanding between employer and employee." In 1922 he concluded a well-publicized ceremony for presenting the first 118 safety committeemen with their badges by calling on the president and general manager to extend to each worker "the open hand of fellowship . . . and to recognize in each the brother bound to you by the most sacred of ties, the bond of noble service, the bond of a common ideal, the ideal of making life more liveable in industry"; and then he asked the committeemen who were about to receive their insignia from these officials to recognize in each of them "a friend and fellow man, just like yourselves, with the same love of family and little ones which you possess, the same emotions, the same aspirations. . . ."[106] This was good theatre but it was also good public relations. The *Sault Star* ended a laudatory article on Algoma's safety work with the question: "Who said Corporations have no soul?"[107]

The steel companies also made gestures toward giving their workers a greater sense of participation in company policy-making, without conceding any important power. The departmental safety committees were a step in this direction, but there were more ambitious schemes. In 1918 Algoma organized a "Welfare Board," which had no "executive powers" but which advised the corporation on "matters relating to welfare and recreation, and to the maintenance of good relations between the Corporation and its employees." Eventually the board administered the Steel Plant Club, the *Algoman*, the company canteen, and the recreation and educational programs. In addition, the firm created a short-lived "Plant Committee," apparently modelled on Mackenzie King's "Colorado Plan" for industrial councils. According to the general manager, this body had "no other jurisdiction than to discuss wages, hours and working conditions." Below it functioned a network of departmental grievance committees. Like other Canadian firms with industrial councils, Algoma found this structure useful in smoothing in wage reductions early in 1921.[108] Both Scotia and Disco proposed councils modelled on Bethlehem Steel's plan at the end of the war, but their workers refused to abandon their demand for collective bargaining through their unions. Only in 1923, after breaking the Sydney steelworkers' union, did the new Besco conglomerate succeed in installing an "Employees' Representation Plan" in Nova Scotia steel plants.[109] Stelco studied the new industrial councils carefully at the end of the war, and early in 1920 called meetings in each department where company officials offered to confer with a committee of Stelco workers. Probably because of the weaker threat from its workers' unions, however, it never actually implemented more than safety committees until the return of industrial unionism in 1935 convinced the corporation to set up a plant council.[110] As in the case of Algoma's Welfare Board, the powers of these councils and the range of issues taken up by their members were quite limited. As the *Besco Bulletin*'s survey of the first two years of that council's operations indicated, however, this restricted form of "industrial democracy" managed to integrate the various strands of corporate welfarism:

> Apart from individual cases, or those affecting one or two committees, the Committees have interested themselves in welfare work and safety campaigns, and have assisted the Superintendents in their endeavours to assure a fair distribution of work during slack and idle periods. Special committees have got thoroughly into the matter of cost with the Management, and have been taken into the confidence of the Company in the arrangement of operating programmes. During the Christmas season, Christmas Cheer was dispensed through the medium of representatives to needy families who would otherwise be forgotten.[111]

All of these attempts to encourage steelworkers to integrate their long-term interests with those of their employers had their parallels in the United States, where most of the programs were first developed. But the particular situation of the Canadian steel industry brought into play another factor tending to tighten the bonds between employers and employed. The chronic instability of so much of the industry made many steelworkers, especially in Sydney and Sault Ste. Marie, anxious about the future of their livelihood. The stable core of these men who clung desperately to their jobs in the 1920s and 1930s occasionally sent petitions or joined delegations to Ottawa to beg for better tariff protection or for government spending to stimulate the industry. Right after the war steelworkers, some of them local union leaders, joined in appeals for assistance through government orders to keep the plants open. Several years later the federal Advisory Board on Tariffs and Taxation heard particularly earnest pleas from Algoma's steelworkers requesting some state intervention to overcome the pattern of sporadic employment. Appealing to the same spirit in the face of the return of industrial unionism, Algoma circulated a handbill in its plant in 1937 that asked: "Don't fight this Company but with it, come to Ottawa with us and get a 100,000 ton rail order." Clearly, the dependence of these steelworkers went beyond social mobility and a modicum of social security, to the very existence of their livelihood.[112]

By the 1920s, then, a Canadian steelworker could approach his employer for help in dealing with personal financial crisis. If he maintained a good record as a dedicated, appropriately deferential employee, he could hope to advance to a better job in the plant and, in at least two of the corporations, could plan on a company pension in his old age. After work he could join a baseball team with his workmates, leaf through a company magazine, eat dinner in a company cafeteria, take his wife to company-sponsored dance, or, in one case, attend a class on steelmaking led by a company supervisor. He could point out the dangerous part of his job to a safety committeeman or bring up a grievance with a plant council representative.

But did these programs and services have their desired effect? Did they lead workers to feel a stronger commitment to their employers and to their work and a weaker interest in unions? The impact of welfare capitalism has been the subject of considerable debate in the American literature on twentieth-century working-class history, between those writers who have emphasized workers' cynicism about all this paternalism and those who insist that workers were won over. Among the latter, David Brody has been the most eloquent in his claim that welfarism achieved its goals of cementing workers' allegiance until the Great Depression of the 1930s destroyed its promise and credibility. Sanford M. Jacoby's more recent survey of the "rise and fall" of this branch of management – indeed, of the whole new personnel-management movement that had emerged in North America during World War I – sees it being significantly de-emphasized by the end of the 1920s.[113]

Assessing the long-term consequences of corporate welfarism in the Canadian steel industry is not easy. One of its chief aims had been to stabilize the work force, reduce labour turnover, and increase productivity. The fragmentary evidence that exists is contradictory. A 1924 study of turnover in Canadian steel plants by the Dominion Bureau of Statistics revealed a 91 per cent replacement rate;[114] yet, corporate managers insisted the rate was dropping. We have already seen that a transition was under way in which some steelworkers were beginning to settle down in the 1920s much more than in the past. But explaining that process must take account of the severe depression that lasted throughout the industry until the mid-1920s, the curtailment of immigration during that period, the winding down of other seasonal jobs in construction and agriculture that had often been combined with short spells in a steel plant, and the irregular patterns of employment in two of the three major steel towns, which would have driven away the drifters in any case. Welfarism may have helped some workers to decide to stay, but for many it must have been more a question of clinging desperately to what they had, without recourse to many other options.

The power of welfarism in undermining the appeal of unionism is equally difficult to determine. Certainly, as we will see, it did not keep unionism out of the Sydney, Hamilton, or Sault Ste. Marie plants permanently. Yet, a few of the social welfare measures may have won some workers' good will and support. A Sault Ste. Marie steelworker who travelled south to help the fledgling industrial union at Stelco in the late 1930s later concluded that waiting for a company pension could make workers loyally quiescent. "It meant that those men who had worked in Hamilton for so many years felt that if they joined up with us, the company would cut them right off the pension altogether," he explained. In Stelco's great, decisive confrontation with industrial unionism in the bitter strike of 1946, some 1,000 (mostly long-term) employees showed their loyalty to the corporation by staying inside and organizing themselves into the "Loyal Order of Scabs." In comparing Stelco and its smaller, non-union neighbour, Dominion Foundries and Steel Company (Dofasco), Robert Storey has convincingly argued that a carefully pitched appeal to the same concern with economic security that so often fuelled the drive to industrial unionism in the 1930s and 1940s could help in keeping out unions – in this case, Dofasco's well-publicized profit-sharing plan.[115]

At the same time, however, an American historian, Gerald Zahavi, has usefully suggested, through his study of a large American shoemaking corporation, that corporate welfare might only bring "negotiated loyalty," that is, workers might accept what was offered as a minimum and as "a source of strength in the negotiation of service and wage concessions from the corporation." That perspective can help us understand Canadian steelworkers' participation in one form of welfarism – the industrial council. Within a few years after the creation of such a council in the Sydney steel plant, former union activists got themselves elected and began to pressure Disco for the same sorts of improvements in the terms of their employment that they had pursued through their union. "The by-laws of the council offered considerable in the way of 'democracy,'" one former Disco worker explained, "and elected representatives were promised protection in their pursuit of justice." In the careful words of Canadian labour economist H.A. Logan, the council "developed ambitions to play a role more vital than that earlier intended by the company." Perhaps most galling for the corporation was the council's success in convincing first the local Ministerial Association and then the Social Service Council of Canada, a reformist body of the Protestant churches, to investigate the twelve-hour day in the Sydney plant in 1929. The churchmen's inquiry, headed by the Reverend C.W. Gordon ("Ralph Connor"), ultimately covered the whole industry and seems to have pushed Stelco into adopting eight-hour shifts on January 1, 1930, and to have brought shorter hours in the continuous departments of the other two corporations the next year. The same kind of agitation emerged in Stelco's works council after 1935. Instead of restraining worker dissatisfaction and promoting deferential loyalty, these councils proved to be a bridge to industrial unionism in the 1930s and 1940s.[116]

Any balanced judgement must conclude that there was probably always a constantly shifting continuum in the steelmaking work force between the sycophancy of the "loyalists" and the deeper cynicism and independence of the "rebels," and that the measures that promised some economic security probably worked more successfully at building steelworkers' "consent" to the company's undisputed control than most of the more transparently patronizing programs aimed at boosting loyalty and morale. Since the welfarism that appeared in Canadian steel plants did so little to change basic problems for workers, such as hours of work and wage rates, it must have been hard for most steelworkers to believe some of the high-flown rhetoric surrounding the new programs.

Storey has added a further emphasis that is so often missing from the American literature on the subject: that welfarism was the velvet glove over an iron fist. In 1920 and 1923, Algoma, Stelco, and Disco were not only striking a new posture of benevolence but also breaking steelworkers' strikes and attacking local unions with firings and blacklistings. Welfarism had the best chance of success when "troublemakers" had been driven out and alternatives to complete corporate control seemed far-fetched or dangerous to those steelworkers left behind. Once the crisis had passed, in fact, welfare measures might then be curtailed. By the mid-1920s *Stelco News* and the *Algoman* had disappeared, the Algoma Welfare Board had ceased meeting, the Steel Plant Club had closed its doors, and the flamboyant McGue had been quietly let go. Only in the later 1930s would a new sense of urgency about the threat of unionization bring back a new wave of welfarism to soften the image of repressive corporate power.[117]

In the final analysis, the element of fear and insecurity must remain at the centre of any understanding of early twentieth-century managerial practices in Canada's steel plants. The newer features of corporate labour policies – the bureaucracy of "systematic" management and the gentler touch of welfarism – had their role, but in the day-to-day administration of the plants, it was older methods and incentives that seemed to ensure the fullest utilization of steelworkers' labour power. Behind the distinctive management practice of the steel companies lay a crucially important structural constraint on workers – the uncertainty of work and wages in a frequently overstocked capitalist labour market. This economic insecurity could push many workers to make full use of the piecework system and to respond to incentives, however demeaning or patronizing, that held out the prospect of regular employment and/or advancement into better-paying jobs. On the same insecurity, supervisors could base their threats to discharge and their patronage systems within the plants. For a steelworker, deference to the authoritarianism of these shop-floor despots could have its rewards, but more importantly, it staved off the likely consequences of any resistance – unemployment and poverty for him and his family. These were the most important strands that caught these workers in a web of dependency.

Not surprisingly, then, many steelworkers who committed themselves to the industry toiled in an atmosphere of fear – fear of injury in the heat, dust, and noise of fire-breathing machines and fear of losing their jobs for insubordination. Ultimately, that fear was mixed with a deep resentment. And once the balance of forces in the steel plants shifted, as they would in each of the world wars, it was the workers with the most pride in their work and the most leverage on the shop floor who would provide the spark for igniting that resentment into collective resistance.

4

Resistance

Workers' behaviour on the job is never easy to predict. Corporate planners may lay out master plans for the operations of their factories and establish rules and procedures to govern the flow of production, but there is no guarantee that the workers into whose hands they place the tools and machines will perform their tasks exactly as expected. In the first place, as we have already seen, many of them have a shop-floor savvy that encourages them to make their own decisions about the job to be done, which no white-collar manager can fully appreciate or anticipate. More importantly, however, they bring to their jobs their own set of priorities and concerns, which can diverge quite widely from those of their employers. The corporations may often assume that their workers' wages are simply another cost of production, but the labour power they have purchased for use in their plants is a highly subjective commodity. Workers have an interest in maintaining or improving their income from wages, on which they and their families are usually completely dependent. They want to prevent themselves from being so overworked that they are incapable of returning to their jobs. They also bring along some non-economic, moral concerns about fairness, equity, and human dignity that do not fit on the company balance sheet and that are often nurtured in the working-class communities where they spend their hours off the job. Consequently, while their concern for a regular livelihood may coincide or converge with their bosses' interests in profit maximization, there is plenty of room for regular, ongoing conflict, as workers resist their employers' new initiatives or long-term policies.

Resistance in the workplace can take many forms, both individual and collective, and its most vigorous form, usually called militancy, is rarely present at all times. Workers respond to any opportunities that seem to offer the possibility of important gains for themselves and their families or for their fellow workers in general, but those opportunities may normally be quite limited. In early twentieth-century Canadian steel plants, management policies were designed explicitly to make it difficult for workers to raise concerns that were not in the best interests of the corporations. Confronting these employers with demands for changes in their labour policies required some power and a degree of independence from the control mechanisms of the steel companies, especially from the web of dependency that we encountered in the last chapter. Otherwise, workers would likely leave the company in the hopes of finding better terms of employment elsewhere or settle into an apparently quiescent acceptance of an industrial regime in which they had little or no voice, with varying attitudes of cynicism, fatalism, silent resentment, or calculated sycophancy. Under what circumstances did the systems of corporate control in the steel plants break down sufficiently to allow and encourage workers to resist?

The first generations of Canada's steelworkers responded to their work environment in both individual and collective ways. The more private responses for most of them were initially casual, short-term attachment to their jobs and then, increasingly, integration into the web of longer-term dependency on their employers. Collective resistance, however, appeared early in the history of the steel industry and reached a peak at the end of World War i. The timing of these moments of confrontation and struggle was crucial: usually, workers had an edge in the labour market, and the social and political climate had shifted enough to undermine the legitimacy of the corporations' control. Equally important were the specific segments of the steelmaking work force involved: invariably the core of the resistance came from those workers with the greatest autonomy from their bosses in the workplace, who ironically could be found at both the bottom and the top of the occupational ladder. Finally, we will see that there were distinct patterns of resistance in each of the steel towns that suggest the range of possible responses from workers to similar systems of production.

Passing Through

A large proportion of Canada's first steelworkers had no intention of letting the occupational label stick. They were simply passing through. They developed no steady attachment to the industry and fitted their stint in a steel plant into a series of short-term work spells that could include farm labouring, construction, and, increasingly, machine-tending in other mass-production plants.

Before World War I there were few detailed reports of this form of working-class behaviour, which would eventually become known as "labour turnover." In fact, management critics would later argue that the failure to keep adequate records of turnover was a large part of the corporations' difficulty in dealing with the phenomenon.[1] But by the outbreak of World War I, there was widening public discussion about workers' inclination to move frequently from job to job. The young American labour economist Sumner Slichter conducted some research for the United States Commission on Industrial Relations, which he later expanded into a book-length study of labour turnover in the United States. He concluded that pre-war turnover had averaged 100 per cent in most factories. A study for the U.S. Bureau of Labor by Paul Brissenden and Emil Frankel suggested the rates were probably even higher. Perhaps the most celebrated example was in Henry Ford's new automobile plants in Detroit and Windsor, where by 1913 the turnover rate had reached 370 per cent, as thousands of workers flowed through the firm's new assembly-line jobs. The U.S. Bureau of Labor published figures on an allegedly typical American steel plant that indicated the company had had to replace an average of 118.2 per cent of its work force each year between 1905 and 1910 (a third of whom were returning to the plant after an absence).[2]

There were no comparable studies of Canadian steel plants. But the local press in factory towns like Sydney, Hamilton, and Sault Ste. Marie regularly commented on the comings and goings of the so-called "floaters."[3] In 1910, moreover, the Nova Scotia Commission on Hours of Labour was able to discover that Disco "retained few or none of its employees for any lengthened period of time." The corporation's employment agent informed the commissioners that during 1908, at the depths of an economic slump, 1,071 new men were taken on and 1,508 were re-employed – a total of roughly two-thirds of the work force. As the American studies had revealed, the highest rate of turnover was among the least skilled workers. Emilia Kolchon-Lach's study of Algoma's Italian workers in this period produced similar evidence. Even in the much more stable steel plant at Trenton, twenty to twenty-five men were still quitting every day in 1915, according to Scotia's general manager.[4]

During World War I the problem seemed to reach crisis proportions for the steel companies, since military recruitment and the closing off of European immigration robbed them of their normal reserve army of labour.[5] The results were frequent labour shortages and declining productivity. Without the whip of poverty to discipline them, workers quit or took days off, confident of their ability to find work again soon. "Employment is so easily obtained that workmen change from one occupation to another for no apparent reason," Nova Scotia's factory inspector reported in 1917, "and employers complain that it is impossible to enforce discipline in their factories." He noted that Disco had had to hire 6,000 workers to fill its 4,000 positions during the previous year. That corporation's general manager saw the immigrant workmen as particular problems: "The foreigners in our employ, particularly the Austrians, realizing the scarcity of labour, are not doing a fair day's work. It is extremely difficult to keep them in hand." By 1918 Disco's president was worrying out loud in the Sydney *Post* about "the very serious loss of production" in the plant "even though we have now more names on our payroll than we had some time ago when the output averaged fifty per cent more than it is today." When the Royal Commission on Industrial Relations reached Sydney the following year, Disco's general manager explained that the corporation had to carry about 15 per cent more men on its payroll than were actually working in the plant because "our men don't come out to work."[6]

The same story was unfolding in Hamilton and Sault Ste. Marie. In August, 1916, the Hamilton *Spectator* reported that one of the city's metal-working factories "usually employing 1,500 men had 2,300 men in three months ask to be paid off. In their places 2,100 were taken on, and it will require 200 more men to bring the staff up to its full strength."[7] In a similar vein, Algoma's president wrote to the Prime Minister in 1917: "whilst numbers are short, efficiency is pathetically low. We rate the efficiency of our labour now at about 60 per cent." The following year a federal board of conciliation was appalled to discover that the 1,200 "Austrians" in the corporation's coke-oven and open-hearth departments "receive such good pay that they do not require to work all the time, so they knock off whenever they like, and do not come back until their money is spent, or they are good and ready."[8] It was this kind of behaviour that prompted the federal government to pass its infamous "Anti-Loafing Law" in the spring of 1918, prohibiting idleness and requiring all males aged sixteen to sixty to find gainful employment or risk $100 fines or six months in jail. The victims of the crackdown that followed were almost exclusively European immigrants, who were already feeling the heat of the rising anti-alien agitations. Algoma, Disco, and Stelco were thus able to use the courts to enforce their rules about giving a week's notice before quitting.[9]

What lay behind all the shifting around between jobs that was so characteristic of mass-production industries in the early twentieth century?[10] In the case of steel, workers were often simply reacting to the ups and downs of the industry. Short-term shutdowns and layoffs were a regular feature of the industry, as furnaces needed relining or rolling mills underwent repairs that could last weeks or months at a time. Some parts of the plants would also experience seasonal curtailment, as on the docks and in the stockyards, where a supply of raw materials would often be laid in for the winter.[11] Rather than sit around in unpaid idleness, some workers would leave in search of other work. (Stelco, in fact, used such a shutdown to help break a strike of labourers at its Hamilton plant in 1915.[12]) More serious were the periodic slumps in the industry – 1904, 1907-1909, 1913-1915, and varying periods between the two world wars – that threw hundreds of men out of work and inevitably brought mass departures from the steel towns.[13]

Since the turnover continued (and normally increased) throughout boom periods as well, however, it is necessary to consider how workers' own inclinations and plans motivated their comings and goings. Most of these so-called "floaters" saw their jobs in an instrumental way, that is, as financially rewarding if unpleasant episodes within a personal or family strategy for longer-range survival. These men fell into two loose groupings. The first included the migrant labourers from eastern and southern Europe and Newfoundland, whom we have already met, as well as many Anglo-Canadian farm boys similarly in search of quick cash. In 1915 the president of Nova Scotia Steel – the corporation with by far the most stable work force – discussed this "large floating population" before a government commission: "There is a lot of transient labor – they are here, there and everywhere. Perhaps they only want to work two or three days to get enough money to take them to Halifax, or somewhere else, and . . . we pay off from 10 to 20 a day."[14]

Mixed in with these purposeful "target" migrants, whose work experience was closely related to personal and family goals for an infusion of cash income, was an indefinable number of footloose workers, who have been immortalized by such writers as Jack London and Robert Service. Often known as "boomers," they were young, single, and highly mobile, moving to the new industrial frontiers in search of high wages and adventure, and notoriously brash and independent-minded. For these men a job in Sydney, Trenton, Hamilton, or Sault Ste. Marie was merely a stopover on a continental trek.[15]

A great many of Canada's first steelworkers, then, never had any intention of settling down in their jobs. Whether it was a longer-term commitment to agricultural pursuits on farms and in villages back home or simple wanderlust, they chose consciously to spend relatively little time in steel plants. Yet for many this instrumental attitude must have been mixed with revulsion at the long hours, rigid discipline, and harsh working conditions. The evidence from the World War I period suggests an aversion to the new work environment available in mass-production industries like steel. When given the chance to work as much and as long as they liked in these lucrative jobs, many quit, took days off to visit the race track, and took it easier on the job. Others hopped from job to job in search of something better – even though conditions of factory life were becoming increasingly similar across North American industries. As many contemporaries recognized, labour turnover was a form of protest that the steel companies were concerned to head off with their new welfare programs at the end of the war. All the business journals bristled with articles on how to deal with the grave problem of labour turnover. "More attention is certainly paid to the purchasing of material and the designing of equipment than to the selection of workmen," Disco's superintendent of industrial relations argued in 1920. "This in no small measure contributes to unrest and dissatisfaction among workmen, and consequently is a dominant factor in labor turnover."[16]

Ironically, a frustrated British socialist had reached the same conclusion about turnover and resistance a decade earlier after working in Canada for some time. "If a labouring man working in, say an ironworks, revolts at the conditions of life and labor, well he may clear out and go west, or he may find other work, railroad building or in the mines," he wrote. "The demand for labor, and constantly shifting habits of a large number of workers, are a 'safety valve' which ward off the social revolution."[17] The writer was pinpointing a phenomenon that historians of the resource industries and transcontinental railway construction have frequently emphasized, but which was just as much a part of early twentieth-century factory life. The first two decades of this century in Canada were a period of dramatically rapid economic expansion, which drew in thousands of newcomers searching for opportunity and/or adventure, but who were making no commitment to the industries of the ruggedly new northern Dominion. The "labour market" in which they participated was not simply an approximation of the supply and demand for jobs, but an unstable, unpredictable surging and flowing of humanity across vast stretches of geography, owing as much to subjective human concerns as to the dictates of capitalist rationality.

Shop-Floor Bargaining

Steelworkers who made individual decisions about passing through Canada's steel mills were unlikely to develop a strong consciousness of themselves as a distinct occupational group with collective interests to advance and defend. Yet, among these workers, the counterpoint to individualized "floating" was, in fact, the solidarity of joint action. Such apparently contradictory behaviour had three sources: the determination of the transient migrants to earn as much and as quickly as possible; the testiness of independent-minded skilled tradesmen; and the emergence of an increasingly stable group among the more skilled production workers who were settling into their jobs in the industry. There were two overlapping and interlocking patterns. First, throughout the period under study (and no doubt well beyond), specific work groups based on skill, occupation, or ethnicity engaged in collective resistance to the steel companies' employment policies. This was the direct action of largely informal shop-floor bargaining over a wide range of issues. Second, some skilled and semi-skilled steelworkers wanted a more formal, institutionalized relationship with their bosses to negotiate employment standards. Only during the special conditions of wartime society were they able to succeed in implanting industrial unionism in most of the steel towns. These new unions attempted to integrate shop-floor bargaining into the union cause and to consolidate permanent gains for the workers. For about two years at the end of World War I, most of these unions managed to hold on, but after 1920 they were all crushed by the weight of employer hostility, economic depression, and organizational retreat at the international union headquarters.

Before the war the most vigorous collective resistance to corporate labour policies came from the bottom and the top of the steelmaking work force – the unskilled labourers and the skilled craftsmen and production workers. Table 3 presents the surviving evidence of strikes in the Canadian steel industry between 1900 and

1923 (undoubtedly a serious underreporting, especially of unskilled strikes) and reveals some interesting patterns. Half of the twenty-nine disputes battled out before 1918 involved the supposedly docile labourers, and all of those were aggressive demands on the corporations for concessions, usually higher wages. In the pre-war years, however, the skilled workers, with one exception, all fought defensive battles against changes in corporate policies that were seen as violating some established standards of fairness. Let us consider each group in turn.

The peasant-labourers whom the corporations had specifically recruited proved to be regular thorns in their employers' sides. Reports on the labourers' strike activity in Sault Ste. Marie are sketchy, though they seem to have been involved in the huge riot of Consolidated Lake Superior employees demanding unpaid wages in 1904.[18] The Sydney steel plant managers had to cope with disruptions in the departments employing these workers from the beginning, as did Stelco almost every year for a decade after 1907.[19] The patterns of these unexpected moments of confrontation were generally similar, as the details of two of the more dramatic events reveal. Early in March, 1903, some 200 Italians and seventy-five Hungarians, along with a smattering of Newfoundlanders, marched out of the Disco plant and paraded through the streets under a red flag to protest the rationalizing of labourers' wage rates that would force many to take cuts. They had decided to turn the tables and demand a wage increase. The strikers posted pickets at all the plant gates, allegedly armed with "clubs, sticks, and even axes," to keep out their fellow labourers. Policemen dispersed the crowds with their own clubs but made only one arrest. These tussles continued for four days before the men capitulated to the corporation's intransigence and returned to work. Disco fired the ringleaders and evicted some 100 men from "Shack No. 6," the apparent headquarters of the agitation.[20]

A much larger strike hit the Hamilton steel plant seven years later.[21] Once again the bulk of the workers were Italian, with a sprinkling of Poles and Hungarians. They wanted a new wage scale incorporating a two- to five-cent increase on the existing 15-cent-an-hour base rate for labourers and rejected their employer's offer of one cent an hour on March 22. On the morning of April 1 the men arrived at work as usual, but ten minutes after starting up they threw down their tools and walked out of the corporation's two main plants. The firm estimated between 800 and 900 men were out, while the strikers set the figure at more than 1,200. Picketing

Table 3
Strikes in the Canadian Steel Industry, 1900–1923

Date	Company	Workers	Issue	Result
March, 1900	Stelco*	200 rolling-mill workers	higher wages	defeat
December, 1900	Disco	200 skilled steelworkers	against wage reduction	success
March, 1901	Stelco*	70 labourers	against dismissal of workers	defeat
August, 1901	Disco	150 labourers	higher wages	compromise
April, 1902	Stelco*	65 labourers	higher wages	defeat
October, 1902	Disco	80 tradesmen	managerial innovations	compromise
March, 1903	Disco	275 labourers	higher wages	defeat
August, 1903	Scotia	blast-furnace workers	higher wages	compromise
June, 1904	Disco	whole work force	against wage cuts	defeat
August, 1904	Disco	200 labourers	higher wages	success
April, 1907	Stelco*	200 labourers	higher wages	success
October, 1908	Disco	60 rolling-mill workers	against new system of tonnage rates	defeat
June, 1909	Stelco*	75 labourers	higher wages	defeat
April, 1910	Stelco*	1,000 labourers	higher wages	success
August, 1912	Stelco	500 labourers	higher wages	success
November, 1912–January, 1913	Algoma	200 tradesmen	against reduction in overtime rates	defeat
May, 1913	Stelco	75 labourers	higher wages	success
July, 1915	Stelco	100 labourers	higher wages	success
July, 1915	Stelco	250 shell workers	higher wages	success
July, 1915	Algoma	300 skilled steelworkers	higher wages, against bullying supervisor	partial success
August, 1915	Scotia (NG-T)†	whole work force	higher wages, eight-hour day	success
February, 1916	Stelco	385 shell workers	higher wages, against bonus system	defeat

Date	Company	Workers	Issue	Result
June, 1916	Stelco	machinists	nine-hour day	defeat
July, 1916	Stelco	150 labourers	higher wages	defeat
August, 1916	Algoma	railwaymen	higher wages	success
January, 1917	Algoma	44 gas-engine workers	eight-hour day	success
March, 1917	Scotia (NG-T)†	200 workers ("repairmen")	higher wages	defeat
May, 1917	Scotia (NG-T)†	60 munitions workers	against wage cut	compromise
May, 1917	Stelco	40 labourers	higher wages	success
March, 1918	Scotia (SM)**	whole work force	union recognition	mere show of force
April, 1918	Scotia (NG-T)†	whole work force	union recognition, higher wages	royal commission appointed
June, 1918	Scotia (NG-T)†	whole work force	same	compromise
July-August, 1918	Disco	whole work force	union recognition	compromise
March, 1919	Scotia (NG-T)†	rolling mill workers	against increased work load	defeat
April-June, 1920	Disco	150 machinists	higher wages, nine-hour day	defeat
April, 1920	Disco	labourers	higher wages	unknown
May, 1920	Stelco	stationary engineers	higher wages, shorter hours	defeat
July-September, 1920	Stelco (Mont.)‡	all rolling-mill workers	higher wages	defeat
August, 1920	Disco	25 firemen	higher wages	defeat
November, 1920	Algoma	200 rolling-mill workers	against increased work load	defeat
November, 1920-December, 1921	Disco	150 railwaymen	higher wages, eight-hour day	defeat
November, 1920	Disco	200 rolling-mill workers	against increased work load	defeat
June-December, 1922	Disco	16 railway workers	sympathy with strikers	defeat

Date	Company	Workers	Issue	Result
February, 1923	Disco	whole work force	against discharge of unionist	success
June-August, 1923	Disco	whole work force	union recognition, higher wages, eight-hour day	defeat

*Hamilton Steel and Iron Company.   †NG-T = New Glasgow-Trenton.
**SM = Sydney Mines.   ‡Mont. = Montreal.

SOURCES: *Labour Gazette*; Public Archives of Canada, RG27 (Department of Labour Records), Strikes and Lockout Files; daily press.

was boisterous but generally orderly despite the presence of policemen. Some of the strikers helped to dump slag and to bank the blast furnaces so that no damage would be done to the equipment. The corporation had initially refused to discuss any grievance or to take back any of the strikers. But on the afternoon of the second day the general manager met a delegation of foremen and convinced them to lead the strikers back to work while a committee discussed their concerns. The corporation had agreed to collect information on wage rates in the Buffalo steel industry and to fire two foremen who had been accused of extorting money from the immigrant labourers for the privilege of working in their gangs. Ten days later the firm announced an elaborate new wage scale with increases of 5 to 10 per cent. In this short-lived burst of militancy – the most serious disruption of the company's production in the whole half-century before the 1946 steel strike – these unskilled workers had effectively challenged, though not eliminated, the steel company's authoritarian, low-wage policies.

There are a number of important similarities between these incidents that characterized most of the confrontations between immigrant labourers and the steel companies. First, no strike of poorly paid workers without valued skills could last long, and all of these disputes were over within a few days. Second, in contrast to many of the craftsmen's concerns, these strikers' demands invariably focused on wages, quite simply because that was their only reason for being in a steel plant. The faster they could earn their nest egg the better. And they were often successful in squeezing an extra penny or two per hour out of the steel companies. Third, each of these events appeared in the local press as a spontaneous upsurge, but they actually revealed a good deal of careful organization. The timing was always wisely chosen to take full advantage of a labour shortage (in the early spring or late summer), and parades and picket lines were normally orderly, despite the hint of violence in the mass pickets and primitive weapons that were sometimes present.

What accounts for this kind of militancy among the element of the steel work force with the least skill and the least familiarity with a steel plant? In the first place, they had a labour-market advantage not available to most steelworkers. As we saw in Chapter Three, the steelmaking corporations had to compete with other employers for these workers in external labour markets. Consequently, when labour shortages developed, as they did frequently before and during the war, the labourers found themselves with a leverage that was not normally available to steelworkers in the mainstream of production, who were caught on the job ladders of their employers' internal labour market. Second, the logic of their "sojourning" in North America could promote militancy: English-speaking workers may have believed that the newcomers would work for any wage, but in fact the labourers wanted to earn as much and as fast as possible in order to return home with the all-important cash savings.

In addition, we should not assume that their various peasant cultures had left them completely ignorant and unprepared for confrontations in their new work worlds in Canada. While it is certainly true that peasant consciousness tended to be limited by loyalty to kinsmen and fellow villagers (*campanilismo* in Italian), most studies of peasant societies have found plenty of evidence of collective action, animated by a deeply ingrained sense of injustice.[22] In the tiny, close-knit ethnic enclaves of the steel towns, ethnic solidarities could survive and be rallied to the fight against employers. The ethnic mutual benefit societies and other institutions in the immigrant communities could often provide some leadership. The small size of each ethnic group in each immigrant ghetto – often no more than a few houses on one street – probably also encouraged more off-the-job contact with other Europeans than was typical in the larger American cities and may have facilitated common action. Among the widely travelled migrants in these communities, moreover, could be found men who had encountered strikes, unions, and radical politics in a variety of industrial contexts on both sides of the Atlantic. The earliest Hungarian association in central Canada, for example, the First Hungarian Sick Benefit Society of

Hamilton, formed in 1907, included among its founders some social-
ists with experience in the United States. Similarly, a Hamilton
worker claimed that the Spanish anarchist who assassinated Spain's
prime minister in 1912 had boarded with him and worked at Stelco
for several months. Canada's flourishing pre-war Ukrainian socialist movement must also have had adherents in the steel towns. It is
not surprising, then, to learn that in 1903 the Sydney press discovered the existence of some sort of union among the labourers
(although it provided no details), and that in Hamilton in 1910 the
corporation blamed the walkout on "foreign agitators, so-called
interpreters who do not work." Clearly, transiency did not necessarily mean passivity.[23]

It is equally significant, however, that the great ethnic gulf in the
steelmaking work force was seldom bridged. "No union exists
among the foreigners except the union of brotherhood," reported
the Hamilton *Spectator* in 1910, "and they realize that the union
does not include English-speaking employees of the steel works, for
so far as can be learned no attempts were made to get them to
strike." The Hamilton *Times* noted during the strike that throughout the immigrant districts "little groups of foreigners were assembled, apparently discussing the situation, but directly a Britisher
hove in sight not a word was heard until he had passed out of
hearing distance." The firms' English-speaking workmen almost
never showed much interest in joining these struggles anyway. If
their normal fears and suspicions were not enough, they were bom-
barded with alarmist press reports about gun-toting strikers, like
an interview in the *Times* with a Catholic priest who warned "that
if they are aroused they will resort to riot without hesitancy, and
. . . that the authorities do not know fully what class of men they
are dealing with."[24] The battles of the European labourers in Canada's early steel plants thus remained as much sectional struggles as
those of the skilled workers.

Ironically, the sources of collective strength that the unskilled
labourers could muster were quite similar to those of the craftsmen
in each firm's mechanical department. Both groups were part of
larger labour markets outside the confines of one company, and
both brought to their jobs independent subcultures of solidarity –
one based on the close bonds of ethnic identity, the other on the
hallowed traditions and practices of the craft and its union. As in
so many other industrial contexts in this period, craftsmen who
still exercised some discretion and control in the labour process
took every opportunity to organize themselves into craft unions to

preserve their workplace customs, status, and power. These men insisted on their right to independence in setting the work routines of their jobs and were outraged by the centralism and bureaucratization of the new factory management practices of the early twentieth century.[25] All the major steel companies had to face occasional strikes by these skilled craftsmen. Although they had united with other practitioners of their trade in the district to establish common terms of employment, their strikes in the steel plants were invariably attempts to preserve some customary work routines against corporate aggression. In the fall of 1902, for example, Disco's whole staff of skilled metal workers walked out to protest the firing of a popular foreman, who had been accused of "not doing his duties," and to get back a paid lunch hour; they were only partially successful.[26]

The flavour of these confrontations between craftsmen and corporations is probably best conveyed in a lockout of some 200 metal workers at Algoma in 1912. The Algoma Iron Works had been a small, general-purpose foundry owned by Algoma's parent, the Lake Superior Corporation. In September, 1912, formal ownership was transferred to the newly reorganized Algoma Steel Corporation, and the company's operations were moved to new buildings within the Algoma steel-plant complex. At the end of the month the workmen who had made this move were informed that they would now have to conform to Algoma's general policies governing wages and hours of work. Specifically, this announcement meant that the workers would lose their guarantee of time and a half for overtime and their limitation on too frequent Sunday work that they had long ago convinced their former managers to accept. Now they would be expected to show up regularly on Sunday and to labour without a full overtime premium. On the first Sunday the workers, many of whom were members of the moulders' and machinists' unions, refused to show up without an assurance that only emergency work would be done and proper overtime rates would be paid. The next morning the company fired the lot of them and refused to discuss the matter with their committee.

The men's statement to the federal Department of Labour rings with the spirit of the proud, independent-minded craftworker, who believed he had well-established, hard-won rights in the workplace. The corporation's unilateral action had violated the customary process of negotiation over these important issues: "If there were any changes to take place concerning time or pay, we feel we should have been consulted, or at least notified," the men argued. Their ethical code was also affronted by Algoma's ungentlemanly arrogance and highhandedness in dismissing them so abruptly, which, they believed, "was unfair and should be beneath the dignity of a great Corporation." Finally, they defended the control mechanism of overtime pay that they had developed to prevent overwork: "the wear and tear of 7 days work a week on our bodies and minds is such that we ought to get extra pay for it. . . . We feel that the time and a half is needed as protection to us from Sunday work." Algoma was clearly unwilling to allow this kind of negotiation over the terms of the workers' employment. It wanted one policy to apply to the whole plant and insisted on its right to adjust that policy unilaterally. The corporation therefore spurned the mediation efforts of local politicians and clergymen and ultimately compelled the men to drift away or to return on management's terms. The craftsmen were brought to heel.[27]

Occasionally groups of workers more centrally involved in the steel production process marched off the job to resist some managerial initiative. The American workmen brought to Hamilton in 1900 to operate the two small rolling mills in the first steel plant there seem to have brought their American union membership with them and early that year walked out on strike, insisting on being paid the new American sliding-scale wage rate. The company refused the demand and held out until some of the men returned to work and others could be imported from the United States. The same year 200 skilled steelworkers in Sydney struck for three days over a wage cut before reaching some undisclosed "amicable settlement," and in 1903 blast-furnace workers at Scotia's Ferrona works used a strike to get higher wages. In 1908 a large group of Disco's rolling-mill hands walked out in protest over the introduction of new tonnage rates, and the men at Scotia's finishing mills in Trenton met the same year to plan a response to a wage cut. Nothing came of either of these initiatives, and, like most of the other pre-war episodes of industrial conflict, these workers' concerns remained confined to specific occupational and ethnic groups within the steel plants. Wartime employment conditions, however, allowed the skilled production workers to fight their sectional battles more aggressively. In July, 1915, 300 workers in Algoma's open-hearth and blooming-mill departments struck to regain the wage rates that had been cut the previous year and to get some relief from an apparently bullying general superintendent. No union was involved besides the bricklayers' organization. After a week the men settled for the promise of a wage increase in the near future and the resignation of the offensive supervisor. Over the next eighteen months the corporation had to make similar concessions to striking railwaymen and gas-engine workers.[28]

Until about the middle of World War I, then, collective working-class resistance to Canadian steelmaking corporations most often took the form of localized battles of labourers, craftsmen, or skilled production workers in the departmental, occupational, or ethnic contexts in which the men worked and rarely involved permanent workplace organizations. By the early years of the war, however, skilled men in some of the steel plants were hoping to bridge these divisions with all-inclusive industrial unions that could promote and defend the interests of all categories of steelworkers. By that point, the idea of uniting all steelworkers in one plant into a single organization had had only one good test before the war – in Nova Scotia at the turn of the century.

Solidarity

The burst of prosperity that began in the late 1890s had spurred on new labour organizing throughout Canada,[29] but it was the unique character of the Nova Scotia labour movement that made possible the first experiment in industrial unionism in the steel industry. Beginnning in 1879 as a miners' union, the Provincial Workmen's Association (PWA) had expanded its membership base into a few other industries by the turn of the century. This was the beginning of a long tradition in the province, especially in Cape Breton, of well-organized, class-conscious coal miners extending the hand of solidarity to other workers in the region. In 1897 the PWA briefly recruited the steelworkers of Scotia's Trenton and Ferrona plants, who were outraged at a wage cut, but they soon abandoned the union in the face of company threats to fire any unionists.[30] The PWA had more success among Disco's steelworkers, who organized two lodges in Sydney in 1902-1903. In October, 1903, the association had no success in talking the faltering corporation out of its abrupt decision to lay off 250 skilled workers and cut all wages by 10 to $33^1/_3$ per cent.[31] But the next spring, the PWA demanded the restoration of the old wage scale, especially in the light of comparative wage data gathered from American steel towns, which showed a 10-15 per cent gap. The union rejected Disco's offer to submit the matter to arbitration and, on June 1, 1,500 PWA members walked out on strike, along with the unorganized labourers and the unionized locomotive engineers, bricklayers, and machinists in the plant.

The PWA lodges quickly signed up hundreds of new members and maintained an orderly strike organization with patrols to curb any violence and a "prosecutor" to check any drunkenness. All the PWA's Cape Breton lodges pledged their support for the steelworkers and opened a $100,000 strike fund. Early in July, 1,000 miners marched through the streets of Sydney to demonstrate their support. The corporation announced at the start that it intended to start up production again as soon as enough workers could be rounded up and warned strikers not to obstruct their passage into the plant. Picket lines numbering in the hundreds were soon turning back the handful of scabs who attempted to enter the plant, though there was no serious violence involved beyond some pushing and shoving. Disco promptly demanded increased police protection from the local authorities. When this proved inadequate to keep the plant gates open, the corporation convinced the local government to call in the militia to disperse the mass pickets. Some of the immigrant labourers were prepared to stand their ground with wooden clubs, but the PWA leadership advised the strikers not to resist.

Instead, they asked the federal government to appoint a royal commission as it had done in British Columbia the previous year, but they succeeded in getting only the mediation services of the deputy minister of labour, Mackenzie King. By the time King arrived in mid-July, however, the militia's work had guaranteed Disco a sufficient work force to get most departments running. The coal miners' threat to strike in support had collapsed in the face of cutbacks and layoffs in their own industry. King's role became simply to preside over the official winding up of the strike on July 22, which involved his secretly agreeing to the blacklisting of twenty-eight union leaders. Henceforth, the corporation announced, "the management will receive only committees of the Company's own employees in connection with any question arising on the plant." The first experiment in using an industrial union to promote and defend the general interests of Sydney's steelworkers had been crushed.[32]

No comparable initiative appeared for another decade. Once again, the first step was taken in Nova Scotia, this time in the Trenton plants of Nova Scotia Steel, which had always prided itself on its good relations with its employees. "Both men and officials grew up with the plant," a local labour leader later admitted, "and in general, there has been a feeling of friendliness between the two and a remarkable freedom from trouble."[33] Traditions were under attack in the corporation's plants, however. The older patterns of work had been shaken up in the half decade before the war, as new technology had allowed for the speeding up and intensification of work on the rolling mills, and the workers had had to accept two wage cuts, in 1908 and 1913. Perhaps even more disruptive was the opening of the Eastern Car Works, Scotia's subsidiary for producing railway cars and equipment, next door to the steel plant in 1912. This new operation brought many newcomers to the town, including American managers and Trenton's first continental European immigrants. The constant resetting of piece rates for new orders in car plants made them a scene of frequent labour turmoil, and Eastern Car was no exception. The tensions in the plants increased as Scotia shifted over to munitions production early in 1915. That spring the resentments of the men in the two plants began to converge.[34]

The new car works had attracted one other important element to the town – a tough-minded, Australian-born socialist boilermaker by the name of Clifford C. Dane, who quickly became the driving force and inspiration for a resurgent labour movement in Pictou County. In November, 1913, the car workers had met to form a union, and Dane became its secretary. The structure they chose, presumably on his advice, was a "federal labour union," a structure created by the American Federation of Labor for unskilled workers who could not be fitted into any craft union. In this case, however, Dane and his fellow unionists transformed the structure into an industrial union. By the summer of 1914 the new union, locally known simply as the Federation of Labor, claimed over 200 members and was maintaining a "Federation Hall" for meetings and socializing. Sometime that year Dane was fired and blacklisted and subsequently became the union's full-time business agent and the *bête noire* of Scotia's management.[35]

Under Dane's leadership, the Federation began to put pressure on Scotia to address the growing list of grievances of the car workers and, by early 1915, of the steelworkers as well. Dane's tactic was to write to the federal Department of Labour to request a board of conciliation under the Industrial Disputes Investigation Act – a device Canadian steelworkers would turn to increasingly during the war years. In February his efforts forced the corporation to restore the 25 per cent of workers' wages that had been cut the previous year at both plants, and in March to adjust the wage scale of some munitions workers. In June the Federation was emboldened by fuller employment and Scotia's concerns about its war contracts to demand an eight-hour day and threatened to pull out all the union men in both plants. The strike plans apparently fell through, but, when the waterboys in the car plant decided to strike for an eight-hour day with no cut in pay and a group of shell workers discovered a new piece-rate scale that could mean wage cuts, all the accumulated grievances and resentments snowballed into a walkout of all workers at both plants on August 16. Three days later they returned to work with the guarantee that their wage concerns would be placed before a federal board of conciliation whose decision would be binding on both parties. When the board convened in New Glasgow late in September, Dane marshalled a parade of witnesses who voiced concerns about long hours, difficult working conditions, and wages that failed to keep up with the rising cost of living or to recognize the workers' indispensable skills in the work process. Their testimony was printed verbatim in the local press and later circulated in pamphlet form. Never before had the working conditions of the community's leading employer been exposed to such intense public scrutiny. The board's final report went some distance toward meeting the workers' demands for wage increases, and the Federation of Labor could claim a victory.[36]

There were problems buried in this first successful episode of industrial unionism, however. Dane and the other committed unionists had been able only to consolidate the many narrow grievances of individual departments and work groups, not to unite all the workers around collective, unifying demands. The union had won considerable respect among many local workers, but the habit of unionism among Scotia's workers was apparently still weak. Whether from fear, inexperience, or indifference, however, they stopped wearing their union buttons, and the Federation of Labor's strength in New Glasgow-Trenton slipped away. Strikes in Scotia's plants in the spring of 1917 followed the pre-war pattern of unorganized battles in single departments.[37]

Dane's office nonetheless became the centre of a whirlwind of organizing activity in Pictou County over the next three years, as he helped various groups of workers organize and initiated branches of the Federation of Labor in local coal towns. Late in 1916 he also helped get an AFL federal labour union off the ground in Scotia's primary plant in Sydney Mines. Then in November a large meeting in Sydney, called with the support of the local miners' union, launched an organizing drive among Disco's workers, which quickly drew in 1,200 members. The Federation of Labor enthusiasts, however, lost control of this project in Sydney when the president of the

Amalgamated Association of Iron, Steel, and Tin Workers arrived on the scene the next month.[38]

The Amalgamated was an old international union that traced its history back nearly half a century. The Amalgamated was originally a craft union, whose strength had lain in the wrought-iron industry, where labour-intensive production processes had required plenty of manual skill and metallurgical expertise. In the two decades before the war, the union was able to hold on only in those plants where new, more automated machinery operated by less skilled men had not yet been developed – especially the sheet-metal and tin mills – or where the enterprise and its market were too small to make these kinds of technological innovations economically feasible. By the middle of the war, however, the union was taking a new lease on life with new organizing initiatives to include virtually all steelworkers in one plant (although deference was still paid to the jurisdiction of the metal-working craft unions, like the moulders and machinists). This was the organization, then, to which Sydney steelworkers decided to hitch their star at the end of 1917, apparently in the hopes of gaining some bargaining clout from the international connection.[39]

The next year was a tumultuous one in the Nova Scotia steel industry. The independent organizations in Pictou County and Cape Breton strengthened their membership base and threatened to bring production in the plants of both Scotia and Disco to a halt. By the early spring of 1918 Dane had spearheaded the organization of seven branches of the Federation of Labor in Pictou County, claiming some 5,000 members in local factories and mines, which met together on April 8 with unionized blacksmiths and electricians to demand that Nova Scotia Steel, Eastern Car, and two small munitions plants recognize their union and grant a 30-cent-per-hour minimum wage. In contrast to the 1915 strike, these demands reflected greater unity and solidarity among the workers. When they were ignored, the steelworkers and metal workers struck a week later, and Federation of Labor miners in neighbouring Westville walked out briefly in support.

Negotiations between unions and management were proceeding simultaneously in Cape Breton and had already included a one-day walkout at Scotia's Sydney Mines plant on March 2. A month later leaders of Cape Breton's two steelworkers' unions joined a large meeting of the island's miners to "map out a plan of campaign for the labor men who are demanding that the corporations recognize the various unions here." Rather than allowing the situation to escalate into a full-scale tie-up of the province's steel industry, with possible support from coal miners, the federal government brought the Trenton-New Glasgow strike to an end by promising to appoint a royal commission to investigate the entire Nova Scotian steel-making industry.

The commission's report was released on May 22 and immediately met a torrent of criticism from Dane and his fellow unionists. Not only were the recommended wage increases for Scotia's workers considered too small – a fact for which Dane roundly denounced the labour representative on the commission, Cape Breton miners' leader James B. McLachlan – but the only concession to the demand for union recognition was a recommendation that the corporation agree to meet "committees of workmen" to discuss grievances. Consequently, on June 13 the Scotia and Eastern Car workers struck again for their original demands. With vigorous pressure not to disrupt the war effort at a crucial moment, the men were convinced four days later to accept the company's new offer of slightly higher wages and a willingness to deal with a committee of employees (but not the union *per se*). Meanwhile, there was renewed fear that the Cape Breton steelworkers would turn the Scotia strike into an industry-wide walkout. In Sydney the Amalgamated lodge was similarly unhappy, but since a federal order-in-council had just urged workers to avoid strikes for the duration of the war, the union hit on the novel tactic of taking a "holiday" each Sunday in July until the Disco workers' demands were met – a tactic borrowed from the miners. These actions were called off when Disco agreed to the appointment of a board of conciliation, and modest wage increases followed in the fall. By the spring of 1919, when the Royal Commission on Industrial Relations arrived in the Nova Scotia steel towns, workers from both corporations could report that informal collective bargaining had been operating for several months. The Amalgamated was co-operating with the craft unions in the plant and had even helped Disco's clerical staff to organize their own union.[40]

Organizing in Sault Ste. Marie had been following a roughly similar, if less dramatic, course since the beginning of the war. In January, 1917, the company's general manager was appalled to learn from his spies that his employees had organized a lodge of the Amalgamated Association of Iron, Steel, and Tin Workers and quickly enrolled some 600 members. By the end of the year the secretary had issued 1,700 cards. Recognizing the advantage that a tight labour market and the urgent demand for Algoma's steel gave them, the steelworkers' union prepared over the summer of 1917 to begin formal collective bargaining with Algoma by working out a new wage scale. The corporation, however, refused to meet with any union representatives, and the steelworkers had to resort to a federal board of conciliation, which the corporation also refused to acknowledge formally. The board adjourned its hearings when the union agreed to give the new general manager, a personable administrator whom the company had plucked out of the army to return to the plant, some more time to review wage rates. By December the union indicated that it was satisfied with the new arrangements. A new board of conciliation had to be convened in the fall of 1918 to settle new wage demands, but as in Nova Scotia an informal dialogue with the corporation seems to have survived into 1919.[41]

The organization of Stelco's plants in this same period was quite uneven. Lodges of the Amalgamated Association emerged in most of the corporation's finishing plants outside Hamilton, which in most cases were antiquated enough to require highly skilled men. A lodge first created by Gananoque's metal workers in 1914 became the centre of a remarkable experiment in community-wide industrial unionism in the months immediately after the Armistice, bargaining through a central committee with all the town's manufacturers. In March, 1919, one of that town's union leaders helped launch an Amalgamated lodge in Montreal's rolling-mill plants, including the three belonging to Stelco. These unionists in turn helped Stelco's rolling-mill hands and horseshoe makers in Belleville to establish a lodge a few months later. Both these unions soon claimed to have signed up virtually all the men in their plants. The most difficult of the corporation's plants to bring into the Amalgamated fold proved to be in Hamilton. After an unsuccessful try in 1918, unionists from Gananoque and Sydney who were in town for the Trades and Labor Congress convention in September, 1919, were able to act as midwives at the birth of a new Hamilton lodge. Membership would eventually reach 1,300, but, as we will see, this union's impact never approached that of the organizations in Nova Scotia and Sault Ste. Marie.[42]

The middle of World War I, then, had been a turning point in the patterns of workers' resistance in most Canadian steel plants. For the first time, workers in all four major steelmaking corporations had made concerted efforts to create new, all-inclusive unions in order to make a more permanent imprint on their working environment. A few striking characteristics of this development should be noted. The timing of this new unionizing activity is certainly note-worthy. It was not until the third year of the war that most of the unions took hold. Full employment at higher wages than ever before in living memory gave steelworkers (and many others) a new confidence based on their greater independence from their employers – reflected, as we have seen, in higher rates of turnover, absenteeism, and reduced "effort" on the job. Equally important, the appeals of jingoism and self-sacrifice had worn thin, and the dislocations of wartime society, especially political mismanagement and private-sector profiteering, were by that point in the war fuelling a growing resentment among Canadian workers.

In fact, the war provided a context for generalizing workers' concerns from the narrower confines of skill, occupation, and ethnicity. As more and more workers began to turn their thoughts to what life would be like after the war, their unions became the vehicle for consolidating wartime gains and ensuring that the clock would not be turned back. "If we don't do something we will get our heads taken off after this great war is over," a Gananoque unionist warned in 1918. "We should take our stand while the door is open for you can see in every-day life that the upper class in every way will clip a laboring man's hair if they get the chance." A Sydney union leader had similar thoughts: "After the war is over, and we have a surplus of labor in the market, is not the time to get justice. . . ."[43]

Second, it is important to note that it was the more skilled workers at the centre of the steelmaking process, especially the skilled and semi-skilled workers on the rolling mills, who provided the cutting edge of this more organized resistance. In Hamilton the backbone of the organization was the group of new men brought from the United States the year before to operate the new sheet-metal mills, and, as the *Amalgamated Journal* reports over the lodge's first year indicated, they were most successful in recruiting other rolling-mill workers at Stelco and neighbouring Hamilton plants. In fact, outside the major primary steel plants (and the Gananoque and Montreal mills), the Amalgamated lodges were mostly craft unions of the skilled men on the more old-fashioned, labour-intensive rolling mills, many of them recent American immigrants.[44] As we saw in Chapter Two, these men had retained intense work-group solidarity and plenty of on-the-job independence from their employers' direct control, and they were now able to translate that strategic advantage into collective power.

Third, the organizers and leaders of this new unionism were all

English-speaking, as they had been in Sydney in 1902-1904, but for the first time they were reaching out to draw in more of the less skilled Europeans. There was still plenty of lingering ambivalence on both sides. In 1919, for example, the Royal Commission on Industrial Relations asked a Sydney unionist whether the "foreigners" in the plant were union members. "Some are and some are not," he replied. "You cannot call them union men because we may see them this week at a meeting and not see them again for 2 or 3 months."[45] Yet there seemed to be a new willingness to recognize the need for cross-ethnic solidarity if steelworkers were going to be able to stand up to their employers successfully. Translating these noble sentiments into action was not easy, especially since the Amalgamated lodges' weekly meetings, whose elaborate rituals and ceremonies closely resembled the gatherings of fraternal societies and craft unions, were probably not the warm cultural events for the Europeans that they were for their Anglo-Canadian workmates. Immigrants with limited language skills, for example, must have been mystified by the codes issued to decipher the new password for that particular season. The unionists nonetheless made an effort to bridge the cultural gulf. Organizers who spoke European languages carried the message to the immigrant ghettoes, and reports began to roll in about success in recruiting the "foreign element." In Sydney they would eventually have their own lodge based in the town's multicultural enclave, Whitney Pier. The experience of the 1904 Sydney strike (and, as we will see, of the tumultuous 1923 dispute) suggests that it was the direct action of the picket lines, not the more cautious, bureaucratic style of most union leaders, that inspired the Europeans' most vigorous support for unions. Had the Canadian steel industry been convulsed by an industry-wide strike comparable to that spreading through the American industry in 1919, they would no doubt have shown themselves to be as aggressive, committed participants as their counterparts south of the border.[46]

A fourth feature of this new burst of unionization was its unevenness across the industry. Although all four major steel centres had new unions by 1919, the relative strength of each varied considerably. For a full explanation we must look outside the labour process to compare the resources available to steelworkers in the communities where they lived and worked. The centrality of the steel plants in the essentially single-industry towns of Trenton, Sydney, and Sault Ste. Marie contrasted with the situation in Hamilton, where Stelco shared the limelight with several other large manufacturing concerns. Focusing community sentiment was thus more difficult in Hamilton. More important, however, were the traditions and institutions of working-class life in each town. Here the contrast between Nova Scotia and southern Ontario is sharpest. In Cape Breton and Pictou County, the coal miners had a history of organization and militancy running back four decades. Organized on an industrial basis, they had already helped the steelworkers in their first effort to create a union at the turn of the century. During the war they had joined Clifford Dane's pan-industry unionism on the mainland, and on the island the miners had provided organizational and tactical support for the new steelworkers' unions. The threat of joint strike action by miners and steelworkers was a powerful club to carry into negotiations with Disco and Scotia – one eventually wielded in the climactic confrontation of 1923. Working-class Hamilton, however, had no comparable source of strength and solidarity to offer Stelco's steelworkers. The city's craft union movement had never had much sympathy with industrial unionism and had to be prodded into helping to organize Stelco in 1919. Craft unionists had no secure base in any of Hamilton's new mass-production plants and had suffered a major defeat in 1916, when Stelco led the city's metal-working firms in crushing a huge machinists' strike. Working-class solidarity thus remained weaker and more fragmented in Hamilton than in industrial Nova Scotia, where AFL-style craft unionism had never made the same inroads.[47] It is clear that similar production processes did not necessarily induce the same responses from Canadian steelworkers.

In the same vein, the decentralized, community-based nature of this new industrial unionism should be emphasized. The organizations in each of the towns were connected to an old, established international union (Scotia's workers eventually affiliated in 1919), occasionally welcomed international officers for brief visits, and sent delegates to the annual conventions. Yet, except for the centrally determined wage scale for skilled puddlers and rollers, the Canadian lodges were left on their own to develop local union policies and strategies. Significantly, they were not drawn into the huge organizing campaign that culminated in the nation-wide steel strike in the United States in 1919. In fact, the eagerness of Canadian steelworkers for all-inclusive industrial unionism was an expression of local sentiment, not a reflection of official international-union policy, which remained much more cautious. Furthermore, there was no clear co-ordination of bargaining strategies within the industry. A Canadian vice-president travelled between the steel towns carrying news and offering advice, and the more committed members could keep up with general Canadian developments by reading the weekly columns in the *Amalgamated Journal*. But for the most part each lodge went its own way.

Finally, the role of the Canadian state in the development of industrial unionism in the steel industry is also noteworthy. During the first three years of the war, the federal government made no significant gestures toward organized labour. Both Prime Minister Robert Borden and Imperial Munitions Board Chairman Joseph Flavelle had resisted the Canadian labour movement's demand for fair-wage clauses in munitions contracts and had co-operated in the defeat of the huge Hamilton machinists' strike in 1916. Only early in 1918 did the government arrange to include prominent labour leaders in some consultative committees on the war effort. No institution ever appeared within the Canadian wartime state wielding the interventionist powers of the National War Labor Board in the United States, which had the authority to recognize unions and impose collective bargaining on reluctant employers.[48] The Canadian government did, however, impose a regulation on Canadian workers that backfired and became an unintended lever for union organizing.

In the spring of 1916 the Industrial Disputes Investigation Act, hitherto applicable only to resource, transportation, and utilities industries, was extended to the munitions industry. Steelworkers and other metal workers were now required to submit their demands to a board of conciliation before striking. The legislation was a decade old by this point and had won the scorn of many of the workers covered by it for the unfairness of the "cooling-out" period of conciliation. Its introduction in wartime became a mixed blessing for Canadian factory workers. The Ontario regional labour paper, the *Industrial Banner*, denounced the measure as an effort "not to protect the workers and remedy the evils they complain of, but to force them to remain at work even if . . . they are refused fair wages and working conditions." On at least two occasions in 1917, the act was used to drag Algoma workers into the local magistrate's court for striking illegally. Yet, for workers in manufacturing who had never been able to get an employer to deal with them, the legislation eventually became a boon. Now industrialists found themselves having to defend themselves to a third party, sometimes in public, against their workers' charges of inadequate wages and unfair employment policies. By the end of the war, Canada's steel-worker unions were regularly requesting boards of conciliation (and occasionally royal commissions) in the hopes of swinging some public pressure behind their demands. In the heady atmosphere of public debate and political controversy about corporate behaviour in Canadian society, they could often expect some support. Significantly, once the post-war crisis had passed, the steel-making corporations refused to have anything to do with this state conciliation machinery. For a brief period, however, their workers had been able to make some limited use of legislation originally designed to placate and restrain them. The state, like the workplace, had become a battleground of class interests.[49]

Toward Industrial Democracy

What did these newly organized steelworkers hope to do with the industrial unions they were struggling to create? Certainly they were looking beyond the specific grievances in the ongoing struggles of shop-floor bargaining. In some ways their aspirations resembled those of their skilled workmates who were clinging to craft unionism. Indeed, this period at the end of World War I was marked by a convergence of labour concerns to create an unusually unified working-class agenda for industry and politics. Workers across the country were insisting on greater access to power at many levels of society, especially the workplace and the state, and for a consolidation of the higher standard of living that many of them had won by the end of the war. They were talking about a reconstruction of society, not simply the settling of some wartime grievances.[50] Yet the particular experience of mass-production workers like those in Canada's steel mills shaped their demands in special ways that should be noted carefully. By examining the writings of their spokesmen, their statements in the press and before commissions and boards, and the demands they placed before their employers, we can piece together the often indistinct yet powerful vision of "industrial democracy" that inspired them.[51]

Among working people with no other source of income besides their wages, the size of their pay packets was inevitably a central concern. One of the first tasks each new steelworkers' lodge set itself was drawing up a new wage schedule to lay before the respective corporations. Wage struggles have seldom been as "pure and simple" as an earlier generation of American writers generally assumed. It is always important to unravel what lies behind demands for more money. First and foremost, organized steelworkers were trying to consolidate the material gains that many had made in their living standards in the wartime boom, but that inflation was seriously threatening, and to spread these new gains to all levels of the work force. Quite simply, in the words of a Sydney steelworker, they wanted "to maintain themselves and their families decently." There is seldom a social consensus about what constitutes a "decent," respectable lifestyle, but steelworkers clearly felt the small progress they had made being eaten away by the final year of the war. Equally disturbing was that prices did not drop much in the two years after the war.[52] The cost-of-living statistics collected by the federal Department of Labour at this time indicate that the workers had good cause for concern, as the retail price curve tilted sharply upward after 1916 (see Table 4).

Table 4
Indices of Weekly Family Budget of Staple Foods, Fuels and Lighting, and Rent in Canadian Steelmaking Centres, 1919–1923
(1910=100)

| | Sydney | | New Glasgow | | Hamilton | | Sault Ste. Marie | |
	Food	Total	Food	Total	Food	Total	Food	Total
1913	116.4	105.4	101.5	101.6	111.4	115.8	108.5	97.2
1914	119.9	103.2	111.6	102.8	110.8	110.1	104.2	97.5
1915	130.0	112.9	117.9	106.9	120.9	117.8	114.3	100.9
1916	160.6	129.7	141.0	122.4	145.3	133.8	140.0	103.0
1917	189.9	148.4	179.8	143.6	178.8	164.2	178.6	144.7
1918	216.2	168.2	203.8	170.5	194.8	183.6	199.0	163.3
1919	227.5	169.4	205.0	186.0	204.9	191.2	203.6	179.8
1920	237.3	185.2	220.0	206.9	208.7	205.1	220.2	190.7
1921	173.4	145.9	160.9	172.3	157.8	176.7	151.6	159.1
1922	163.1	139.9	147.4	163.1	150.5	174.6	146.8	144.6
1923	170.3	143.8	157.3	170.1	159.0	178.4	151.4	141.9

SOURCE: These indices are derived from my own calculations of weekly expenditures, based on the *Labour Gazette*'s monthly retail price statistics for each industrial community (the December reports are used here) and a slightly modified version of the Department of Labour's family "shopping basket," used to compute its national cost-of-living index each month.

This concern with maintaining living standards lay behind much of the outcry against the use of "alien" labour during and following the war. "Cheap alien labor must not be allowed to encroach on the hours and wages of good Anglo-Saxon union wage earners," cried New Glasgow's labour paper, the *Eastern Federationist*. "Let not the enemy come in and take our birth right," Sydney's labour paper insisted. Behind this impassioned language lay not so much an irrational racist hatred as a fear of how immigrant labour was typically used to degrade living standards: "they met the cost of living and saved billions . . . because they live on the cheapest kind of food, wear the coarsest, cheapest kind of clothing, and live in shacks and huts, just what our work people would have to do to save billions. Is there any person in Canada who would have them get down to that system of living?"[53]

Besides their concerns about a decent living standard, many of Canada's first steelworkers also believed that they were not earning a "fair" wage. Inevitably these notions of natural justice involved points of comparison. Often they charged that the rewards for their labour were unfair in the light of soaring corporate profits.[54] They also measured their earnings alongside those of other groups of workers, objecting to unequal wages for the same jobs in other Canadian steel towns or in the United States and to the wide range of wage rates for the same work in one department, which resulted from the arbitrariness of front-line supervisors.[55] Moreover, as we saw in Chapter Two, the more skilled workers in the mainstream of the production process at Scotia's Trenton plant in 1915 wanted wage increases that recognized their indispensability and brought them up to the earning power of machinists and other recognized craftsmen. By the end of the war, many workers in the industry were making similar claims, as the wartime boom skewed the customary ratio between the earnings of different groups of workers. They expressed some resentment that the hefty wage packets that got so much press attention during the heydey of munitions production had gone home in the pockets of the most transient elements in the work force, while the steady, committed workers in the industry had never had access to such lucrative work. "The majority of the men on the plant at present did not share in these high wages," Trenton's steelworkers told a board of conciliation in 1920, "because they were kept in their old positions of skill and responsibility in order that the whole plant might continue to operate successfully."[56] That resentment blew up into a fury when the high wages were flowing into the pockets of the less skilled eastern Europeans in the plants.

It was also significant, however, that large spread in wage rates was also considered unfair. For the first time, the more skilled men were prepared to help bring up the wage rates of the less skilled proportionally faster than their own. The steelworkers' lodges frequently demanded minimum wage rates for the least skilled and higher increases for those earning less than what they considered a living wage. In the spring of 1918, for example, Sydney's steelworkers were agitating for a four-dollar-per-day minimum wage, and by July they were using their weekly Sunday "holidays" to extract from Disco a 15-20 per cent increase for the lower job categories (the great majority) and only 6-8 per cent for the highest paid. Another round of wage negotiations in October was similarly weighted toward the lower-paid men. A year and a half later, Scotia's Trenton workers were making similar demands.[57] These initiatives revealed the power of industrial unionism to challenge the individualism and competitiveness built into the corporations' managerial policies. They also suggest the remarkable degree of working-class solidarity that had emerged by the end of the war.

The steelworkers wanted not only fair and decent wages; they also wanted to establish some independence from total corporate control on the job. That meant primarily the right to negotiate the terms of their employment through some kind of collective bargaining arrangement. That right also implied some greater respect for their indispensability as producers in the steel industry. Unquestionably the ideological climate of the war had helped to shape this new desire. The relentless reminders of their importance to the country's industrial life and to the war effort, which had been used to stimulate their productivity, had sunk in. "We today as men have a standing in this community that should not be underestimated by any of us," wrote an Algoma worker in 1917. "What the union men feel is that as they are chiefly employed in making munitions for our country at war, they occupy just as important a position in feeding the guns at the front as do the men who serve the guns," another worker insisted a few months later.[58] More than anything else, however, it was the pervasive wartime rhetoric of democracy and public service that strengthened the resolve of these workers to challenge the conditions of their subordination in Canadian capitalist society. The fight for democracy abroad brought the promise of real democracy at home, including *industrial* democracy. "There is no use in standing against democracy, for what is democracy, but standing together for the right of each to the fullest proceeds of his or her labor?" the Sydney labour paper insisted. "An underpaid democracy held down to the freedom of contract strictly conditioned by bargaining on the lowest level of subsistence is no democracy at all."[59] A Sydney steelworker expressed the defiance that these workers had discovered: "The time is past when any concern, it matters not how big they are, can say, You must work under the conditions we say or not at all. The worker has learned that to run the mills they must have the men, and they will claim the right to say what conditions they will work under."[60] These workers could also draw upon the same potent wartime rhetoric to denounce their employers' resistance. Disco's general superintendent won the title of "the local Kaiser," and the corporation's unwillingness to accept the employees' right to negotiate was labelled "Steel-gloved Kaiserism." "Does the Company realize that it has at last gone too far and that the people of Cape Breton have suffered too dearly on the fields of Flanders and at home to be ruled by an autocracy?" John Gillis's paper asked when Disco fired the president of its clerks' union.[61]

Here, then, were men who had made a commitment to their industry and who recognized that they still played a vital role in the production process despite the great changes in the labour process. And now they could use the negotiating strength of a tight labour market and the new ideological climate to consolidate some effective power in a more democratic workplace. For the leading spokesmen of the steelworkers' union lodges, democracy in the workplace generally did not mean domination through full-fledged workers' control. Clifford Dane flamboyantly donned the mantle of a "Bolshevist" before the Royal Commission on Industrial Relations in the spring of 1919,[62] but most union leaders at this point emphasized simply the right to negotiate a broad range of issues with their employers. At Disco, Scotia, and Algoma, the workers proposed a formally structured committee of union and corporation representatives, which would meet at regular intervals to discuss wages and working conditions. In none of the big steel centres, however, was full-fledged collective bargaining ever developed to the satisfaction of the men. At various times between 1918 and 1920, Disco, Scotia, and Algoma seemed to be willing to discuss grievances with a central committee from each lodge, and even, quietly and reluctantly, to ignore the fact that a full-time local union leader might be part of these delegations. In the summer of 1919 the Sydney lodge actually convinced Disco to accept the "check-off" (deducting union dues from workers' wages and turning them over to the lodge),[63] but generally, the corporations avoided as much as possible tying themselves down to any formal arrangements. In fact, as we saw, they soon began to press their workers to accept some form of industrial council. The union men rejected these plans unless, like Britain's "Whitley Councils," the new structures incorporated union representatives.

They were just as resistant to other forms of corporate welfarism. Collective bargaining, they insisted, implied the preservation of workers' independence from the corporations through their unions, so that their particular interests could not be submerged within corporate concerns about profitability. When the Social Service Council of Canada sent a delegation to Sydney in 1919 to investigate employment conditions in the steel industry, they found the steelworkers "unwilling to receive on a charity basis anything whatever from the company. They want justice and if given it they will find their own club houses and recreation; they will not have benevolence from the company until they are given justice as human beings."[64] The following year the Sydney lodge made a direct assault on Disco's welfare programs by getting four unionists elected to the board of directors of the firm's mutual benefit society. The Sydney Mines union made a parallel but unsuccessful effort to scuttle Scotia's benefit society in favour of union representation.[65]

For these mass-production workers, workplace democracy also did not imply the formal labour-market and job control that craft unionists demanded – apprenticeship regulations, workload limitations, and so on.[66] Steelworkers had no leverage over the external labour market, since, as we have seen, they were generally promoted through the ranks within a single firm. But they did want to curb the excesses of the men who controlled access to the higher rungs on those job ladders – the foremen and superintendents who still held the effective power to hire, promote, and fire. The Algoma lodge's 1918 proposals for union recognition and collective bargaining emphasized the need for "reasonable discipline" and for a formal grievance procedure to deal with "small or petty grievances" arising on the shop floor. Here was the direct link with the existing shop-floor bargaining. The same concern with arbitrary favouritism had prompted the PWA lodges at Disco and the later Amalgamated lodges to demand a system of promotion by seniority. These attempts to bring some fairness to the administration of the labour process would become a central feature of most mass-production industrial unionism – what one writer later called "civil rights in industry, that is, of requiring that management be conducted by rule rather than by arbitrary decision . . . a system of 'industrial jurisprudence.' "[67]

The question of job control is more complicated. Unlike craftsmen (including coal miners), most steelworkers had never known a world of small-scale production by self-employed artisans. However much they may have recognized and insisted on their skills in the steelmaking process, it would have been hard for them to imagine such large-scale, capital-intensive factories without a distinct management (capitalist or otherwise). Their criticisms of corporate officials often amounted to charges that they were incompetent managers, and even in the most militant steel towns in Nova Scotia, union leaders had praise for men with good managerial abilities.[68] Certainly none of their formally drafted demands on the steel companies proposed the negotiated control mechanisms that craft unions regularly insisted upon. Yet, since formal collective bargaining was so limited, we have no definite answer to the question of how far Canada's steelworkers were willing to push into their employers' unilateral control over the labour process. The issues they raised before government commissions and in some of their grievances suggest that there was little that was beyond the pale. After explaining the evils of the piece-work system, a Trenton union leader indicated that it was "the aim of our organization to do away with the system and have hourly wages, which will give us a decent living wage in return for which we purpose [sic] giving an honest day's labor."[69]

In a similar vein, rolling-mill employees at Disco, Scotia, and Algoma demonstrated by striking that they thought they had the right to negotiate the size of the work group on each mill. Scotia's workers in particular wanted to put pressure on the corporation to use some of their large wartime profits to modernize their production facilities in order to relieve some of the hard manual labour required on much of the firm's antiquated equipment.[70] At the National Industrial Conference in the fall of 1919, Sydney's John Gillis declared that workers wanted an end to "long hours of labour and other speeding-up processes."[71] These would be the so-called "management rights" that would be removed from collective bargaining when industrial unionism finally took hold in the steel industry in the 1940s.[72] But in this early phase, unionized steelworkers seemed to want an openness and flexibility to co-determine many aspects of the labour process in which they participated.

Only in Nova Scotia did the leaders of the steelworkers' unions question the legitimacy of corporate ownership and control. Here the battle lines were drawn between the local community and insensitive "outsiders." After Scotia was taken over by American interests in 1917, the steelworkers of New Glasgow-Trenton and Sydney

Mines were outraged at the new owners and their managers. Disco's absentee owners prompted even more anger. In 1918 a steelworker wrote to the Sydney *Post* to complain that Disco officials "neglected to study the temperament of the people they had to deal with," whose proud independence led them to expect that any demand they raised "should be theirs by right and not as a concession granted by any corporation." That resentment against arrogant outside control would reach new heights after the creation of the new consortium, the British Empire Steel Corporation (Besco), in 1920.[73]

The steelworkers' new notion of "democracy" was not limited to the workplace. In each of the steel towns, their union leaders were active in broader working-class efforts to ensure a greater voice in the political life of the country, most particularly the various local Independent Labor parties.[74] In the 1917 federal election, Sydney steelworkers' leader, John Gillis, ran as an ILP candidate in Cape Breton, and in 1920 a former Scotia machinist, Forman Waye, won a seat in the provincial legislature for the party. In the same election the New Glasgow-Trenton workers had put up their union president, H.D. Fraser, but without success. Several Labour aldermen were nonetheless elected in the Nova Scotia steel towns. In Sault Ste. Marie an Algoma employee, W.T. Whytall, headed up the local ILP, and three to four steelworkers made up the bulk of the Labour aldermen elected between 1918 and 1921. In Hamilton, steelworkers organized late in the development of the city's triumphant ILP, but one of Stelco's skilled rolling-mill hands, Bert Furey, was active in both the Hamilton branch and the province-wide ILP.[75]

At the same time, among many European steelworkers, two new ideological strains were inspiring the search for industrial democracy. One was nationalism. For some immigrant groups, organizing to support the war effort in Canada had helped to unify their normally fragmented local communities and to inspire a new patriotism. For eastern Europeans the surge of enthusiasm for national self-determination following the breakup of the old empires further galvanized the "foreign colonies" in towns like Sydney and Hamilton. The latter's Polish community published a statement early in 1919 declaring their unity "in their burning and heart and desire to see Poland restored to her old glory in her unity, freedom, and independence." The second new factor was revolution. While Labourism was the dominant political expression among the English-speaking union leadership, the Russian Revolution had inspired much more radical politics in the immigrant ghettoes, especially among some eastern Europeans. One of Stelco's spies reported in the summer of 1919 that Hamilton's European immigrants with whom he came into contact "appeared to be thoroughly poisoned with Bolsheviki propaganda." The Hamilton branch of the One Big Union had many of these "Bolshevists" among its several hundred members. It was undoubtedly the oratory of these radicals at the steelworkers' organizational meetings in the city a year later that led a visiting organizer to express surprise that "the soviet idea was so rampant here among the foreign-speaking iron and steel workers."[76] Repressive state policies that included arrests and deportations forced most of these workers into a cautious silence, however, and they ultimately had little direct impact on the policies of the new unions in their industry.[77]

The aspirations of Canada's steelworkers for decency, fairness, and democracy converged on the most important demand of the post-war Canadian labour movement, and the one with the most immediacy for the steel industry – the eight-hour day. This issue crystallized the confrontation between humanity and profits. Shorter hours could be at least a partial solution to the cruel fate of unemployment. They were also essential to creating more fairness and democracy in Canadian society by allowing steelworkers the opportunity to enjoy the full life as citizens and family members.[78] In fact, this was an aggressive demand not simply for opening up more leisure time but for limiting the impact of the new pressures of the work world on workers' non-work time, for creating more space in human activity for independent activity free of the control of corporate employers.[79] Arguably, this was a union demand that connected most directly with the individualized resistance to steel-plant work routines that were producing the staggering rates of labour turnover.

Of course, steelworkers were not alone in these concerns. They were inspired by the wider working-class struggle for shorter hours and the wide-ranging post-war debate about the issue. "They believe from statements made by politicians and writings in the daily press, and all that – they feel they are entitled to an 8-hour day," one steelworker explained to the Royal Commission on Industrial Relations. In Cape Breton the miners' success in winning an eight-hour day in February, 1919, gave the local steelworkers an immediate point of comparison, and at mass meetings on the island in the spring and summer of that year they expressed their resentment at the inequality.

Several experiments with shorter hours in the Canadian steel plants also proved disruptive by giving workers a taste of something better. Early in 1915 Scotia had put some of its munitions workers on eight-hour shifts, and by the summer other groups of the corporation's workers, including the waterboys, were demanding the same treatment. In August, 1919, Algoma's blooming-mill and rail-mill workers resisted the return of the twelve-hour shift after a few months on eight-hour shifts, as a result of reduced production. They were evidently attempting to use the occasion to win the eight-hour day permanently, but they lost. The next year, the men on the merchant mill proposed three eight-hour shifts as an alternative to an announced reduction of the work force on the twelve-hour shifts, but they, too, were ignored. The boldest initiative came in Trenton in the spring of 1920, when Scotia's workers simply began to organize their own new weekly work schedule with Saturday afternoon off, "until such time as the eight-hour day comes into effect." Their protest was suspended for a board of conciliation investigation and then succumbed to the curtailment of production.[80]

The Amalgamated lodges (and, of course, the craft unions in the industry) tried at various points to negotiate shorter hours with their individual employers, but invariably they encountered their employers' arguments that the competitive conditions of the industry prevented them from acting unilaterally. Increasingly, therefore, the steelworkers' union leaders turned to the state for legislation to bring the corporations to heel. In the spring of 1919 Nova Scotia's steelworkers joined with other unionists in demanding that the provincial government enact an eight-hour measure, under the threat of a general strike. Under strong counter-pressure by the local branch of the Canadian Manufacturers' Association, the province dodged the issue by agreeing to appoint a commission of inquiry into the implications of shorter hours.[81] In Ontario, the Algoma lodge president, Erie Dalrymple, attempted unsuccessfully to convince Premier William Hearst, the Sault's MLA, of the necessity for such legislation.[82] Far more attention was directed to winning over the large National Industrial Conference convened in Ottawa by the federal government in the fall of 1919 to thrash out post-war labour policies. Sydney's John Gillis was among the labour speakers who filled the House of Commons chamber with eloquent, impassioned appeals for relief from the rigours of ten and twelve hours of daily labour. The corporate magnates sitting opposite them, however, were unmoved and again recited their belief that Canadian industry could not support such a fundamental change as long as other countries still clung to the longer working day.[83]

Over the next few years the representatives of capital and labour carried these debates to the conventions of the new International Labour Organization, which soon endorsed the eight-hour day for all industrial countries. But the Canadian business community still resisted, arguing that the ILO's resolution was being ignored in most countries.[84] The initiative eventually returned to the provincial legislatures. In Nova Scotia the ILP MLAS were unable to convince the house to adopt an eight-hour bill, and in Ontario the ILP cabinet ministers had no more success with their coalition partners in the Farmer-Labour government.[85] As with so many other parts of the steelworkers' vision of "industrial democracy," reducing the burden of their daily toil would have to await another day.

What was going on in Canadian steel plants (and in steelmaking communities) by 1919, then, was a contest for power. Workers were not necessarily out to win complete control, but they wanted an extension of their own power to allow a sharing of control that would enable them to pursue their own class interests. They were not all contesting their subordination within monopoly capitalist society, but they were certainly challenging the terms of that subordination.

Confrontation and Defeat

Canadian steelworkers had emerged from the tumultuous war years determined to entrench and formalize their fragile new collective bargaining arrangements and to use this new institutionalized power to improve their terms and conditions of employment. They knew the post-war equilibrium had not yet been established in the workplace: as Forman Waye told the Royal Commission on Industrial Relations: "They recognize our union when we hold the whip hand, when we are strong enough to warrant it; but when we are not they do not recognize it."[86] By the fall of 1919 the whole American steel industry was convulsed by a national strike that the steelmasters seemed determined to break.[87] Across Canada employers in many industries were locking horns with the local labour movements and provoking general strikes in cities like Amherst, Toronto, and Winnipeg by the spring of 1919.[88] But the struggles were unfolding at different paces in different communities and industries. In steel, the unionists watched anxiously for the shape of their employers' post-war labour policies.[89]

The owners and managers of Canada's steelmaking corporations initially seemed startled and uncertain about how to deal with this new challenge. Like the rest of the Canadian business community and the Canadian state, they were astonished at the depth of disaffection among the country's workers and felt their way carefully and cautiously through the minefield of post-war labour relations.[90] During 1919 and 1920, markets remained unstable, and in most steel towns production was sporadic. The flexibility of their diverse pre-war labour force was declining as returning veterans demanded the dismissal of "aliens" and many of the Europeans headed home. With the lingering menace of an unsympathetic public opinion and the resurgence of a powerful farmers' protest movement (which won control of the Ontario legislature in 1919), corporate leaders initially seemed more concerned to placate the new unions in their plants and to draw them into an alliance to protect and stabilize the industry.[91] By the early months of 1920, however, a consensus had apparently emerged among the corporate leaders of the steel industry to follow the example of their American counterparts, who had broken the great steel strike of 1919. They decided to root out the new steelworkers' union and to find a basis for re-establishing their undisputed hegemony over the steel plants. They were aided by the growing anti-labour momentum of the national "red scare" that was being fanned by the press in the wake of the Winnipeg General Strike and the creation of the One Big Union.[92] The result was a series of skirmishes with the various Amalgamated lodges that reached a violent climax in Sydney in 1923.

Each of the steel companies began by softening up their workers with 10 per cent wage increases in the early months of 1920[93] and by launching the sophisticated new corporate welfare packages that we examined in Chapter Three, including recreation programs, pension plans, safety drives, and company magazines. Before long, however, they were revealing the iron fist inside the velvet glove. The first round involved groups of skilled workers organized outside the Amalgamated, but the main steelworkers' union soon felt the attack. In May, 1920, Stelco stood firm against the demands of its operating engineers for a shorter working day and higher wage rates to maintain their incomes. The strike brought serious disruption of production, but Stelco fired the strikers and imported strike-breakers from its out-of-town plants and from "other outside plants which were indirectly interested in the fight between the engineers' union and the company." These scabs were housed and fed on company property and protected by Thiel Detective Agency staff. The men were forced to return on their employer's terms.[94] Stelco also dug in its heels against its Montreal rolling-mill employees who, along with the city's other millmen, walked out on July 15 for a wage increase. Arguing that the men had not given adequate notice, the corporation withheld their outstanding wages and refused to meet any negotiating committee. For two months the strikers' ranks held firm, but by late September Stelco had coaxed back enough workers to start up the mills again. The union formally called off the strike on September 29 on the condition that all workers would be taken back, but its president found himself blacklisted in all the city's rolling mills. The Amalgamated lodge in Montreal had been destroyed.[95] Meanwhile, Stelco had broken its agreement with its Gananoque workers and refused to have anything to do with the new lodge of Hamilton workers.[96]

In Sydney and Sault Ste. Marie, the corporations were just as vigorously attacking the unions in their midst. Disco managed to break a lengthy machinists' strike for higher wages and shorter hours in the spring and summer of 1920.[97] The defeat of these skilled workers had a severe dampening effect on Sydney's steel-workers. "The failure of the machinists' strike recently has made the majority of the men more careful about walk-outs unless the cause is really serious," the press reported.[98] By the fall, Disco and Scotia were preparing to fend off yet another challenge from their skilled workers, this time the unionized railway workers on their property, who wanted wage parity with other North American railwaymen and an eight-hour day. The corporation faced a brief boycott by the Railway Brotherhoods early in 1921, but it managed to break both the long, bitter strike and the union.[99] In November, 1920, Algoma took the same steps against its most skilled rolling-mill workers, the 200 men on the merchant mill, who struck in November against the corporation's decision to cut one worker off each of the four-man rolling teams. When the strikers decided to return to work after two days to give the new system a trial they were locked out, and their leaders were blacklisted when production started up again. Since some of these men had been key leaders in the Amalgamated's Algoma lodge, the union was severely weakened.[100] The steelmaking corporations would henceforth have no truck or trade with unions.

The material basis of the unions' original bargaining strength was simultaneously being cut out from under them. The winter of 1920-1921 saw the first drastic curtailments of production and layoffs of thousands of steelworkers.[101] The corporations now had a large pool of desperate unemployed workers to draw upon. Early in 1921 all the steel companies felt they now had the freedom to cut wages by an average of 20 per cent. When all the Nova Scotia steelworkers' lodges requested a federal board of conciliation to consider such a reduction, the corporations (now merged into Besco) bluntly refused, knowing that in this economic crisis they held the upper hand. Further cuts followed later in the year.[102] For Canadian steelworkers, as for the rest of the country's workers in the early 1920s, keeping body and soul together became a much more pressing priority than holding together a union organization. Many left the steel towns in search of work. Others accepted almost any terms of employment offered by the steel companies in order to put bread on the table for their families. The optimism and determination that had carried these workers into the post-war period were rapidly dissolving into demoralization, and the unprecedented commitment to a new set of industrial relations based on class solidarity slipped away.

In the difficult new period that began in 1920, Canada's steelworkers were further hampered by the vehicle they had chosen to express that solidarity. The Amalgamated Association had formally opened its ranks to the unskilled and by 1919 had swept up thousands of them in the North American steel industry. Yet in many ways it had never made the complete transition from its earlier days as a craft union. Despite the objections of many local officers, the union's leadership allowed other craft unions to pluck men out of Amalgamated lodges to be merged into city-wide craft locals.[103] Machinists, for example, were encouraged to cultivate their solidarity with skilled metal workers in other plants rather than with the men around them in the steelmaking production process. These skilled workers made separate demands on their employers and even waged their strikes independently.

Common action with Amalgamated lodges could and occasionally did occur, but the union's head office did not encourage these moves – especially after the crushing defeat of such an alliance in the great American steel strike of 1919. That defeat, in fact, brought new caution to the union's leadership. Drawing back to the safe, familiar ground of the skilled rolling-mill worker became the union's new strategy over the next year.[104] In the Canadian lodges this meant carving up the large industrial locals into smaller lodges based on occupational groups within the steel plants. The skilled steelworkers with the strategic advantage in the production process were thus cut off from the workers with less skill and less leverage. In Hamilton the most active group within the lodge, the sheet-mill workers, were shut away in their own union, as were the men in the west-end "Ontario Works" (formerly the Ontario Rolling Mills). The rest of the membership was lumped together in a single, amorphous lodge, which soon expired.[105] The timing of this kind of fragmentation – the summer and fall of 1920 – could not have been worse. In Sault Ste. Marie, for example, an international organizer hived off the merchant-mill workers into their own lodge in the fall of 1920 at the very moment when Algoma was provoking a strike by intensifying the men's work load. Unable to call on a broader membership for support, these workers floundered in a narrowly sectional struggle and were reduced to approaching the city council for help as a mediator. They lost, and the remnants of the Amalgamated in the city disintegrated soon afterward.[106]

At least one voice rose from the Amalgamated's Canadian membership to protest how poorly this resurgent craft unionism was serving Canadian steelworkers. A feisty young worker in Gananoque, who had been at the forefront of building a successful industrial union in that town, was bitter. In the fall of 1922 Gordon Bishop ran unsuccessfully for the position of Canadian vice-president on a platform emphasizing full-scale industrial unionism, in alliance with some "Progressive" militants in the United States, and, in the dying days of the Gananoque lodge, he used his column in the *Amalgamated Journal* to fire a barrage of criticism against union policy that he believed favoured the skilled over the unskilled. "The fight to make our union an industrial one in the full sense of the word has to my mind not been completed," he wrote. Not only were men forced off the union rolls when they could no longer afford the high dues; the remaining members faced additional assessments to replenish the union's coffers. Bishop found these levies discriminatory: "Plainly speaking the 1 per cent assessment is to raise a fund to defend the skilled workmen. Unless the skilled workmen of our organization show a little more consideration towards the unskilled and unorganized there is little hope for advancement."[107]

This combination of employer hostility, economic crisis, and organizational retreat spelled the end of the Amalgamated in Canada. Lodges collapsed across the country in the early 1920s. By 1923 the Amalgamated's Canadian membership was only 135, and the international headquarters decided to abolish the position of Canadian vice-president.[108] But far out on the east coast, a group of Sydney steelworkers was determined to buck the trend and to continue the struggle for the kind of industrial democracy that had captivated the imagination of so many of their fellow workers at the end of the war. The town's Amalgamated lodge had stagnated like the others. An active member remembers that turnouts at meetings had shrunk to about twenty. Craft unions of blacksmiths, moulders, bricklayers, and carpenters had also appeared to fragment the workers.[109] But once again it was Cape Breton's miners who made the difference. During the spring and summer of 1922 the miners had vigorously resisted a wage cut with a prolonged strike that ended in early September.[110] Emboldened by this example, and by the return of somewhat fuller production, the steelworkers' union at Disco began a new organizing drive and even added a new lodge based in Whitney Pier, made up predominantly of less skilled Europeans. That fall the union proposed the revival of collective bargaining with committees of workmen. Disco refused and instead resurrected its plan for an industrial council. In December, however, the corporation's scheme met a humiliating defeat in a referendum of the corporation's steelworkers.[111]

By the early weeks of 1923, membership in the Amalgamated lodge was growing "with leaps and bounds," according to the local press, and the union now laid new demands before the corporation for a 15 per cent wage increase and an eight-hour day. Not surprisingly, Disco rejected the demands, along with a subsequent request for a board of conciliation. Even the Sydney Ministerial Association failed to get the corporation to budge. On February 13 the accumulating tension exploded when an active union member, Sid McNeil, was fired over a disagreement with his supervisor about running wire-making machinery without an operator. Some 275 of his workmates walked out in support when the corporation refused to meet a union committee to discuss his grievance. The strike quickly spread to most of the plant, even though union organization was far from complete. The union insisted on an investigation into McNeil's dismissal and raised a new demand: recognition of the union had to be made a condition of the settlement. Mass pickets shut down the plant, keeping out even maintenance staff and secretaries and blocking any approaching trains, especially the one carrying bedding and supplies for managerial staff at work inside. The press reported that the coal miners' radical leaders were meeting with the Amalgamated leaders to offer "advice and assistance in their management of the details of the strike," and noted the use of the same tightly disciplined form of strike organization developed by the miners the year before. According to Labour MLA Forman Waye, the steelworkers' union secretary, the miners were "ready to take 'drastic action' if the steel strike is not settled within a few days." Disco initially fired off pleas to the federal and provincial governments for military support, but eventually the company agreed to talk with a union committee. An agreement to have McNeil's case investigated brought the strike to an end after four days. The corporation's general superintendent had also made an informal commitment to resume discussions with the union. For a union that had so recently been completely ignored, these were significant concessions. The leadership proclaimed a "glorious victory" over the corporation's policy of discriminating against unionists.[112]

Once production had returned to normal, however, Disco's management wasted no time in returning to its traditional no-nonsense authoritarianism. The official "investigation" of McNeil's case predictably found his dismissal for insubordination valid (he was quietly removed from the scene with a company ticket to the United States). A 10 per cent wage increase announced in March was the only friendly gesture the corporation made that spring. In mid-March it issued a statement once again proclaiming its commitment to the open shop. At the end of the month, it got warrants for the arrest of thirty-five steelworkers for their part in blocking and unloading trains on their way into the plant during the strike.[113]

In fact, Disco was soon able to get the full repressive power of the state on its side in its efforts to break the steelworkers' union. Nova Scotia Premier E.H. Armstrong shared the concern of Disco and its parent conglomerate, Besco, that the militancy of Cape Breton miners and steelworkers had become well mixed with a tough-minded radicalism. Specifically, the influence of the new Workers' Party of Canada on local labour leaders had Armstrong thundering against the "Bolshevik menace" on the island, especially after such prominent central Canadian Communists as Tom Bell, Malcolm Bruce, and Jack Macdonald arrived on the scene. That spring Armstrong authorized the expansion of the hitherto tiny provincial police force by recruiting, according to local lore, "the idlers and ne'er-do-wells who inhabited the rum dives and houses of ill fame along the Halifax waterfront." Some 1,000 of them arrived in Sydney on railway cars bristling with machine guns and were billeted on Disco's grounds. The steelworkers' union was soon complaining to the premier that these "constables" were "improperly disciplined, being allowed to roam around the streets in an intoxicated condition and carrying firearms, thereby jeopardizing the lives and property of peaceful and law-abiding citizens."[114]

Actually, this new force was working closely with Colonel D.A. Noble, Disco's security chief and a Cape Breton military intelligence officer, who had expanded his longstanding internal spy system to root out unionists and to keep the provincial police abreast of developments inside the local labour movement. Furthermore, after perhaps half the steelworkers stayed off work on May 1 to join the May Day demonstrations in Glace Bay, the roving provincial police launched a series of almost daily raids on offices and homes of Cape Breton's union leaders in search of seditious literature. No arrests resulted before the police were sent home in May, but a concerted effort was now under way to portray Cape Breton's union leaders as irresponsible revolutionaries who were leading otherwise contented workmen astray – a theme that the local and provincial press quickly picked up. "Isn't it time the government attempted protection for the safe and sane worker by driving these radicals from Sydney?" the Sydney *Post* screamed in early May. Not since the Winnipeg General Strike had Canada seen such a well–co-ordinated campaign by the state, media, and private industry to destroy a workers' organization.[115]

During these same weeks, the steelworkers' union leaders were working at their organization's membership, fighting individual grievances, and continuing their discussions with the miners about strategies for mutual support. By early June the union leaders were aware that Disco's order book was at last filling up again, and, with that bargaining edge and an overwhelming strike vote from the membership, they approached the corporation for a 20 per cent wage increase and for union recognition through the "check-off" of union dues. The union spokesmen made clear that they would have the full backing of the miners in the event of a strike. The corporation's board of directors, meeting in Montreal, nonetheless bluntly refused these requests, and on June 28 the steelworkers walked out again, completely shutting down the plant. This time there was complete unity across ethnic divisions, as the small European community in Whitney Pier threw itself wholeheartedly into the strike.[116]

This was the ultimate confrontation for which the February strike had been merely a warmup. The Amalgamated's international president, M.F. Tighe, arrived to urge caution and restraint,[117] and the local lodge president asked for restraint on the picket lines. But many Sydney steelworkers were in no mood for politeness. Direct action to prevent Disco from running its plant, they reasoned, would be the only means of having their demands taken seriously. Huge crowds of angry picketers confronted the police at the plant gates with stones, bottles, and insults, as a magistrate tried in vain to read the Riot Act. The next day some strikers invaded parts of the plant to drive out maintenance men but met violent resistance from a force of 400 "loyalists" inside the plant armed with iron bars and led by the corporation's security chief. The next morning the first 250 of an eventual force of 1,500 soldiers in the Canadian militia arrived from Halifax, with the authorization of the local county court judge, and pitched camp on Disco's grounds. And a day later "Armstrong's Army," the provincial police, returned to the scene. The resulting clash was, in the words of historian Don Macgillivray, Cape Breton's "Peterloo." The crowds of picketing strikers now faced a formidable force of military might and soon found themselves staring down the barrels of machine guns and scurrying out of the way of men on horseback wielding clubs and sabres. On Sunday, July 1, the provincial police made a particularly violent assault on an unorganized crowd of unarmed strikers and churchgoers returning from evening service on the main street of Whitney Pier. These displays of armed aggression quickly curbed the mass demonstrations outside the plant, and the provincial police continued to prevent crowds from assembling anywhere in Sydney's streets. There were reports that this armed force was even used to escort strikebreakers to and from the plant. Disco was consequently able to start up production on a modest scale within a week after the strike began.[118]

The outrage of "Bloody Sunday," however, had brought most of Cape Breton's miners out on strike on July 3 in protest against the use of military force to break the steelworkers' strike. "No miner or mine worker can remain at work while this government turns Sydney into a jungle," J.B. McLachlan insisted in a ringing circular to the miners. The long-threatened alliance of miners and steelworkers had finally materialized in the form of an overtly political protest against the state's repressive role in local industrial relations. Working-class outrage at the treatment of the striking steelworkers was spreading westward. Letters protesting the Nova Scotia government's actions poured into Premier Armstrong's office, and a few of Alberta's coal miners joined the eastern sympathy strike.[119]

Repression came swiftly and bluntly. The miners' union president and secretary, Dan Livingston and J.B. McLachlan, were promptly arrested and spirited away in the dead of night to Halifax to stand trial for seditious libel. Several days later the international president of the United Mine Workers of America, John L. Lewis, who had been looking for an issue to dislodge the radical Cape Breton leadership, responded to an appeal from Besco by declaring the miners' sympathetic strike in violation of their existing collective agreement and placing the union in trusteeship. He also ordered the Alberta miners back to work.

By this time Disco was engaged in negotiations with the steel-workers' union, with Premier Armstrong and the Reverend Clarence MacKinnon (a prominent Maritime social-gospeller) as mediators, and, according to the recollections of Doan Curtis, one of the strike leaders, some kind of agreement looking toward an eight-hour day in the continuous-production departments was in sight when a copy of Lewis's telegram arrived. The talks came to an abrupt halt, and by the end of the month the steelworkers' strike and the miners' sympathetic walkout were petering out. The repression was demoralizing enough, but the recent years of unemployment and short time in the steel plant, along with the severe cuts in wages, had left Sydney families with few reserves to hold out for long without wages. Disco heightened this anxiety at the end of July when it announced that strikers would have to vacate company housing. On August 1 a disheartened meeting of 500 unionists accepted their executive's recommendation to call off the strike.[120]

The aftermath had a familiar pattern. Dozens of high-profile union activists were blacklisted and forced to leave Cape Breton in search of work. The union leadership was thus effectively driven away and dispersed. For the strikers taken back into the plant, a retired steelworker later recalled, "the bosses took advantage of you. . . . If you were active they'd give you the dirty end of the stick in any job there, they wouldn't give you justice. Give things to the fellows that weren't active." Another remembered that "they put me in purgatory – put me on the back shift for a year. Gave me a labour job in the billets."[121] At the same time, however, as in the immediate post-war period, Disco made efforts to restore morale and win back the allegiance of their workers. When a group of company "loyalists" approached the corporation immediately after the strike to request some kind of plant council, General Superintendent Bischoff dusted off the "employee representation plan" that he had proposed the previous winter and arranged for elections of worker representatives at the end of August.[122]

Prime Minister Mackenzie King, meanwhile, was appalled that the mechanisms of the Militia Act had drawn his government unwillingly into the great siege of Cape Breton and was determined to prevent a repeat performance. In September he therefore set a royal commission to work to investigate the industrial conflict in Sydney over the preceding many months. Its hearings provided a lopsided view of the events, since, as even the *Financial Post* had to admit, many of the most critical witnesses had already left. Whatever recommendations the commissioners may have had about the Militia Act, they bought the version of recent history promoted by both the corporation and the provincial government – that the whole episode was simply the result of machinations by the "reds."[123]

Industrial unionism soon gave up the ghost in Sydney. The remaining unionists announced in December that they were disgusted with international unionism and were sending back their charter to the Amalgamated. They promised to organize an independent Canadian union but were unable to do so for another twelve years. The One Big Union attempted to reorganize the steelworkers in 1924 but made no headway. The defeat of the strike had made a deep imprint on workers' consciousness in Disco's plant. "They sort of threw a scare into the men, as far as organized labour was concerned, for a long time," one of the future union leaders recalled about his first years in the plant after the strike. "They were scared and when anyone would talk union to them they would rather have you talk to someone else. . . . I used to hear some of the old union men talking, and they were watched very closely. A lot of them were blacklisted a long time before they got back on the plant." Some of those older activists eventually got themselves elected to the plant council, where, as we saw in Chapter Three, they attempted to use that confining structure to pursue some of their old goals, especially higher wages and the eight-hour day. That, however, was a difficult uphill battle with no significant victories, since, as several workers had told the 1923 royal commission, "committee men are more or less in the position of mendicants; they cannot enforce any demands."[124]

An important phase of worker resistance to the terms and conditions of employment in Canada's steel plants thus came to an end in the mid-1920s. The steelworkers' post-war vision of "industrial democracy" had finally been smashed. The steelmaking corporations had vigorously rejected the idea that their workers might exercise some formal power in co-determining workplace policies.

With the active assistance and collaboration of the Canadian state,[125] the corporations had re-established their systems of hierarchical, authoritarian control, which left the workers in a state of fearful dependence and vulnerability. In the rhetoric of the wartime unionists, industrial "Kaiserism" and autocracy had been firmly implanted in the heart of Canadian mass-production industry.

Conclusion

Between the 1890s and the 1930s, working in steel became a new kind of job experience in Canada. Like several other mass-production industries, the steel industry was built on a new system of manufacturing that confronted workers with new machines and work routines and new concentrations of power in the hands of their employers. And, like their counterparts in those other industries, steelworkers sought new ways to pursue their own goals within these new structures.

This new system of production was not implanted easily. The capitalists who launched the Canadian steel industry at the turn of the century faced a host of daunting problems. Their raw materials were often unreliable, and the market for their steel products was uncertain and unstable, as they had to cope with the rigorous competition of foreign (especially American) producers and the gaping holes in the Canadian tariff structure. In the face of these challenges, four major corporations emerged to dominate the industry, each with integrated facilities for mining, processing, and finishing iron and steel, dispersed widely over the Canadian landscape at Sydney, New Glasgow-Trenton, Hamilton, and Sault Ste. Marie. Two of these, Disco and Algoma, were what we might now call "megaprojects" and a third, Scotia, was drawn into the monstrous Nova Scotia conglomerate of the 1920s, Besco. All three faced serious problems by 1920 as a result of internal financial manipulation and heavy dependence on the railway industry for a market. Of the industry's "big four," only Stelco had the diversified production lines and access to wider markets in southern Ontario that could be translated into a relatively stable, successful business. As an industry, Canadian steelmaking remained a set of truncated fragments that never came close to dominating the domestic market in primary and semi-finished steel products. Despite a pre-war and wartime boom, it was an industry over which a sense of crisis seemed to hang for much of the half century before World War II. For steelworkers, this industrial structure would mean fragmentation and economic insecurity, as many of them were made to pay for their employers' questionable investment decisions with a lower standard of living.

Within their factory walls, Canada's first steelmaking corporations attempted to match their American competitors with the most sophisticated production systems available, and in the process helped to inaugurate the Second Industrial Revolution in Canadian manufacturing. Since the directors of the new corporations were generally not experienced iron and steel men, they hired American managers to oversee plants with American production techniques and, initially at least, a considerable number of American workmen. Visitors were astonished at the huge "gigantic automatons" that the corporations had created, where massive machinery carried raw materials and finished products quickly through the stages of production. This new labour process promised greater managerial control and thus cheaper labour costs and cheaper steel. But the reality fell short of corporate expectations. The dazzling array of new "labour-saving" machinery reduced the overall manpower requirements but introduced new demands for skill at the core of the production process and in maintenance work, as well as a growing reliance on the experience of semi-skilled machine-operators. These workers were not simply interchangeable parts in a mechanically controlled system. Many of them were allowed a wide degree of personal independence in shop-floor decision-making, simply because the successful, efficient production of steel relied on the accumulated knowledge of co-operative teams of skilled and semi-skilled workmen. And many of them developed the pride and determination of men who knew their value as the "producers" in the industry, even if their skills had been learned informally on the job and consequently did not give them the independence of old-time craftsmen. The corporations had consequently not escaped their dependence on workers' skill and had to find ways to hold onto this valuable, seasoned work force, without conceding them too much workplace power.

In stocking their plants with workers, the steelmaking corporations made some new departures that brought about a set of occupationally and ethnically segmented labour markets. The most skilled workers were initially recruited from abroad and then, more regularly, drawn from the ranks of longer-term employees in the plant. Eventually a variety of welfare measures were beamed at these more valuable workers in the hopes of encouraging them to settle into their jobs. For the least skilled jobs, the corporations hired Canadian farm boys, but they also followed the American example of reshaping the unskilled labour market by recruiting migrant European and Newfoundland labourers – men without any long-term commitment to the industry, who were willing to work long hours at low wages in the harsh new work environment before returning to their homelands, and who had limited contact with the more skilled Anglo-Celtic segments of the work force during their sojourn in the steel towns. This was a solution to corporate labour needs that seemed to work best before 1920, but as three of the four major corporations slumped into the difficult interwar period, the "birds of passage" were seen less often in Cape Breton and Sault Ste. Marie. They continued to migrate to Hamilton until 1930, when the new depression reduced their job opportunities and immigration restrictions cut them off from their homelands. By that point, the flood of transient workers who had passed through the steel towns had left behind a considerable number of Europeans who had chosen to settle, but overall, by the end of the period under study, they comprised a smaller proportion of the steelmaking work force than they had in the heyday of sojourning, especially in Nova Scotia. The steel companies' recruitment policies had thus ensured that a cohesion between stabilizing, Anglo-Celtic skilled workers and transient, European unskilled workers would be extremely difficult before the 1930s.

Moulding and disciplining this new work force to get the maximum out of their labour power was also a challenge for the steel-plant managers. The new machinery itself could not provide much direct control since so much of it demanded the workers' judgement in operating it. Corporate officials therefore relied in part on the new bureaucratic monitoring of "systematic" management – from cost-accounting to time-clocks – but even more on versions of the familiar incentives of the "carrot" and the "stick." Steelworkers were kept on the job for twelve-hour days, as often as possible paid by group piece-rates to stimulate their self-interest in speed-up, encouraged to compete among themselves, and pushed by gruff, no-nonsense foremen. Even after the creation of central employment offices, these shop-floor despots kept many of the workers under control through their still surprisingly effective power over hiring, firing, promotion, and individual rate-setting. Many workers, especially the Europeans, thus lived in dread of their supervisor's wrath. The corporations heightened that atmosphere of fear by firing and blacklisting union activists and maintaining a network of spies connected to the company police and the local constabulary. Fundamentally, these practices rested on manipulation of the workers' anxiety about unemployment and poverty. When that anxiety lifted appreciably during World War I, the corporations made new, more humanitarian gestures toward their workers with various welfare programs to regain their loyalty, but once the crisis of post-war labour resistance had passed, they turned back to the older forms of repressive management. Ultimately the steelmaking corporations offered their workers a trade-off based on personal self-interest: in return for complete obedience and passive acceptance of corporate control, the workers got the promise of higher earnings through incentive wages and the possibility of economic security through regular employment, promotion possibilities on the firms' internal "job ladders," and perhaps a pension or insurance plan. The personal independence allowed to many of them on the job also left room for workers' quiet pride in their skills and manliness. As a system of workplace discipline and control, it relied much more heavily on authoritarian supervision to keep workers' minds fixed on meeting their material needs in this way and to curb any resistance, than on bureaucratic direction by so-called "scientific" managers.

Unfortunately for the steel-plant managers, many workers found the price of this trade-off too high. Serious working-class resistance to these corporate labour policies consequently did arise, in both individual and collective forms. In the first place, thousands of workers simply passed through the plants, often fitting these brief jobs into a varied work career that could include farming and other resource-extraction jobs, and refusing to settle down and accept the new work world of mass production that had opened up. Partly, this behaviour reflected personal and family strategies for building up a nest egg of cash savings, but it also suggested plenty of disgust with employment conditions in the huge, intimidating, dangerous plants. During World War I this labour turnover became one of the industry's most serious problems, along with absenteeism and reduced "effort" on the job – in short, a massive "refusal to work." This individualized form of working-class resistance would plague steelmaking corporations until the 1920s and 1930s, when workers' choices of alternative employment (especially in construction and resource industries) declined and their free flow across international borders was inhibited by economic depression and immigration restrictions.

Collective resistance did not emerge easily among such footloose workers. It was difficult, in fact, to get much co-operation or solidarity within a work force drawn from such a variety of backgrounds and with such a wide range of expectations. Before World War I, it was rare to find much co-operative action that united the specialized steel-production men (often from American plants), skilled tradesmen, and ethnically diverse unskilled labourers who found themselves gathered under the same large factory roof. Yet, the absence of plant-wide organization did not mean workers never joined in collective resistance. On the contrary, the solidarity of work groups based on occupation or ethnicity frequently erupted into confrontations with the steel-plant management. These were usually episodes of informal shop-floor bargaining over a wide range of corporate labour policies, from low wages to abrupt changes in customary work routines. The most militant of these groups were the least skilled and the most skilled, both of whom had some independence from the control mechanisms of the corporations. Rarely was there any formal organization involved. This was not simply an "immature" form of resistance that would eventually evolve into unionism; it was a regular, endemic pattern of workplace conflict that would continue into the period of unionization and that unions would later try to channel into more formal procedures.

By World War I, however, there were skilled and semi-skilled men in the steel plants who wanted the more permanent, regularized negotiating relationship with their employers that they believed plant-wide industrial unions could offer. They saw the need to impose the kind of bureaucratic restraints on steel-plant managers – a "rule of law" in the workplace – that union contracts could provide. Nova Scotia had provided the only Canadian examples of this kind of organization at the turn of the century, but Scotia and Disco had crushed those early unions. Nova Scotia steelworkers led the way again during World War I. At New Glasgow-Trenton, Sydney Mines, and Sydney, they benefited from the support of the region's tough-minded coal miners and from the relatively weaker impact of exclusivist, mainstream North American craft unionism. By 1919, however, industrial unionism had taken hold in all the Canadian steel towns, as the industry's workers, like so many others in Canada, dug in their heels to protect their wartime gains and ensure more economic security, decency, fairness, and democracy in post-war society.

The diversity of the steelmaking work force made the base of industrial unionism fragile and its accomplishments all the more remarkable. The strength of these new unions rested on the skilled production workers, especially the men in the rolling mills, who had only limited support from the Europeans with less commitment to the industry. An additional organizational problem was the craft unions' insistence on separate units for the craftsmen in the steel plants. Ultimately, in 1920, the steelworkers' own international union, the Amalgamated Association of Iron, Steel, and Tin Workers, would retreat to its craft-unionist core and split up the local lodges in order to protect the skilled rolling-mill men. For some two years, the industrial unionists nevertheless managed to develop the first, fragile collective-bargaining arrangements in the Canadian steel industry.

It took the aggressive resistance of employers during 1920, the debilitating shock of mass unemployment that hit late that year, and the simultaneous organizational retreat of the Amalgamated Association to destroy this struggle for "industrial democracy." All the industrial unions dissolved in the early 1920s, except the militant lodges in Sydney. There the steelworkers were outraged by the behaviour of the region's new corporate monopoly, Besco, and were inspired and encouraged by Cape Breton's determined coal miners and their radical leaders. The final destruction of this phase of industrial unionism in the Canadian steel industry thus became a pitched battle in the streets of Sydney in the summer of 1923, when Disco fought its striking workers with the aid of hundreds of armed militiamen and provincial police. The defeat was brutal and complete, if not final.

For more than a decade, most Canadian steelworkers were not prepared to risk another collective struggle against their bosses. Labour turnover declined, and by the 1930s a more permanent work force had appeared in all the steel centres – one that would be more susceptible to the whims of front-line supervisors who still controlled access to the jobs and promotions within the plants. With a large pool of desperate unemployed workers outside the factory gates, most steelworkers, like so many other Canadian workers, kept their heads down, tried to stay in the good books of their foremen, and looked to personal strategies of survival.

Within these general, industry-wide patterns lay important local variations in working-class resistance, which suggest how the peculiarities of industrial and occupational structure and of working-class community life can produce different outcomes. New Glasgow-Trenton was the oldest steel town in the country and the most unusual among the four largest. After the opening of Scotia's blast-furnace facilities in Sydney Mines in 1904, New Glasgow-Trenton became a centre for the firm's finishing work alone. Its work force was locally recruited, ethnically homogeneous, relatively stable, and generally more skilled as a result of the requirements of the older, more labour-intensive technology. (The same characteristics were found in the small southern Ontario towns where Stelco had finishing plants, notably Gananoque.) Here in Pictou could be found some of the most committed, self-aware steelworkers in the Canadian industry. They had little direct contact with the mainstream North American craft-union movement, but they did have close links with the independent-minded coal miners of Pictou County, who had a long history of militant regional unionism in the Provincial Workmen's Association. The combination of skill, ethnic cohesion, regional identity, and working-class solidarity fired up a remarkable movement of industrial unionism, which was eventually undercut by the curtailment of production in Scotia's plants after the war and the Besco merger.

Sydney, on the other hand, began the twentieth century as a more "typical" North American steel town, with a huge, highly mechanized, narrowly specialized steel plant and an ethnically stratified work force. Gradually, Disco's workers took on a more regional flavour, as outsiders settled in, as local men began moving up the firm's job ladders, as the European migrant population tapered off, and as more regular contact developed with Cape Breton's militant coal miners. Nowhere else in Canada were mass-production factories and coal-mining communities close enough together to create this particular working-class community. Here, as in Pictou County, craft-union exclusivism was weak, and working-class solidarity in the face of outside corporate capitalists produced the most determined unionizing efforts in the Canadian industry. It took the twin blows of brutal repression and prolonged economic crisis to bring these steelworkers to heel.

Like Sydney, Sault Ste. Marie was a company town with a highly specialized, crisis-ridden steel plant at its core. But there were important differences for the city's workers. Far out on the edge of "New Ontario's" wilderness, Algoma's steelworkers were quite isolated from both the fragmenting force of southern Ontario's craft-union movement and the alternative working-class culture of resistance in the Nova Scotia coal towns. They had to rely more on their own internal resources and, by the end of World War I, managed to evolve a cautious brand of industrial unionism. But, with the industry's highest percentage of Europeans in their ranks, they seemed to have faced some of the most severe ethnic divisions of any of the steel towns. The community's isolation, the fragility of Algoma's existence, and the cultural tensions between groups of steelworkers thus put restraints on the Sault workers' ability to assert themselves.

Hamilton's steelworkers lacked the community-wide focus on their workplace that existed in the other "company towns" where steel was produced. Instead they were immersed in a large, diverse working-class population, which allowed for both a vigorous, exclusivist craft-union movement and a considerable degree of ethnic segmentation. By the end of World War I, Stelco's Hamilton workers also faced the most powerful, most successful corporation in the industry, which stood shoulder to shoulder in Hamilton with several other corporate giants. Where Nova Scotia steelworkers could turn for support to nearby coal miners, Stelco could rely on the capitalist solidarity of Westinghouse, International Harvester, and other local corporations in maintaining a union-free environment for capital accumulation. (Similar conditions prevailed in Montreal, where Stelco had important rolling-mill plants.) Greater corporate power and weaker working-class solidarity therefore help to explain the difficulties Stelco's workers had in mounting an effective challenge to their employer's hegemony.

This, then, was how mass production came to Canada – corporate factories, sophisticated new technology, and tough new managerial practices, all of which required the thorough subordination of workers. That subordination was never, in fact, complete, but steelworkers were also never able to create their own successful agenda for industrial politics. It is a story not of the complete and permanent shattering of all workers' on-the-job power and independence, but of corporate employers' largely successful efforts to keep workers from using what workplace influence they still had to make their own demands on the industry. The managers of these corporations wanted to hold on to these workers and were prepared to allow many of them considerable discretionary power in the workplace, provided it was directed to maintaining the rapid flow of good-quality steel. What they would not concede was the same discretion in determining wage rates, hours of labour, and other terms and conditions of employment.

The steelworkers' experience was not unique. Looking back into the nineteenth century, we can see some striking parallels in the work world of Canada's first great capitalist corporations, the railways: brand new technology with new skill requirements filled as much as possible by internal recruitment; segmented labour markets for the less skilled labourers and for the more skilled running trades and shop crafts; welfare programs to stabilize workers and engender loyalty; tough measures to instill discipline; highly bureaucratic managerial procedures combined with the personal, idiosyncratic power of front-line supervisors; and struggles by railway employees to establish greater fairness in the corporations' labour practices.[1]

The conclusions reached in this study have more important implications, however, for a better understanding of work in the first mass-production industries in this country and for the general history of the twentieth-century working class. First, the experience in steel underlines the importance of not abstracting individual labour processes and the factories that housed them from the larger economic structure in which they existed. As we have seen, product markets can have a major impact on the world of work. Many branches of Canadian manufacturing must have suffered the strains of a limited domestic market, as well as the booms and busts of the widely fluctuating business cycles.[2] Workers in Canadian manufacturing towns were particularly vulnerable to these fluctuations since each community had generally become so specialized and so dependent on single industries (in contrast to the more mixed industrial towns of the late nineteenth century).

Second, there were important structural dimensions to the recomposition of the Canadian working class in the early twentieth century. The working class was not born once and for all time in the nineteenth century; its structure changed as a result of the new recruitment and employment policies of monopoly-capitalist industry. In the new mass-production industries, corporations brought together workers from a variety of backgrounds with no experience of living and working together and thus introduced new social and cultural divisions. The argument is frequently made that, after the First Industrial Revolution, workers faced changes in their working lives with less residue of a pre-industrial culture and more industrial experience. Some workers faced the great changes that began in the 1890s with that kind of background, but mass-production industries generally had brand new work processes, often with little direct continuity with the past, and incorporated a large percentage of workers who had not grown up in the shadow of smokestacks.[3] Many Canadian employers, like capitalists in other Western industrial nations in this period,[4] relied on a steady flow of migrant labourers from agrarian settings with no intention of staying here. The ethnic diversification we saw in the steel industry was just as evident in auto production, meatpacking, agricultural implement work, and many other large-scale factory operations.[5] Since it took a long time for settled communities with shared experiences to develop among workers from such varied backgrounds, co-operative action among them was difficult (though by no means impossible). For many years, the first generations in mass production inevitably turned to the particular forms of association and struggle that were most familiar, notably the ethnic solidarities of the European newcomers and the exclusivist craft unionism of the most skilled Anglo-Canadian workers.[6]

Corporate employers also integrated their workers into their factories in different ways. The more valuable workers in the mainstream of the production process became more deeply enmeshed in the firms' internal promotion systems – undoubtedly one of the most important and distinctive workplace changes of the twentieth century – while both the unskilled labourers and the highly skilled tradesmen operated in more independent external labour markets. Workers thus found themselves segmented into distinct slots in the work world on the basis of occupation and ethnicity, most often through informal processes rather than by consciously manipulative managerial strategies of divide and rule. This view calls into question the highly influential recent analysis by David Gordon, Richard Edwards, and Michael Reich, which suggests that "homogenization" was the main theme in the early twentieth-century work world.[7] Any working-class activist from the period would have been more struck by the fragmentation and diffusion of his class. By the 1930s these differences had begun to diminish, as more workers were incorporated into regular jobs within their industries, but it took the collective action of a new generation of industrial unionists in the 1930s and 1940s to build bridges across the divisions that had been created in the working class.[8]

Third, the impact of technological change needs more careful attention than it has generally had in discussions of this early twentieth-century period. Historians and social scientists should not leap to easy assumptions that a new machine equates with a stripping away of all skill. Studies of the First Industrial Revolution have made clear that new skills emerged and old skills survived, and we should approach the Second Industrial Revolution with the same sensitivity to a new range of skills.[9] The transformation of working life in the age of monopoly capitalism was not a one-dimensional process of "deskilling." A few craft skills survived because the new technology required careful maintenance. New groups of skilled production workers also emerged because the machinery and the quality of the product required careful monitoring and control by experienced workers, often in co-operative work teams. The evidence presented here about skill in the steel industry could be tested in other process industries, like chemicals or pulp and paper.[10] Even in the American automobile industry, where the celebrated assembly line has always been assumed to have wiped out all skill, Nelson Lichtenstein has found important pockets of skilled workers still needed in the late 1930s, when no more than 15 to 20 per cent of the auto work force was actually on the "line."[11] In the same vein, we must approach the great new category of "semi-skilled" as something more than thinly disguised labourers who could be treated as interchangeable parts. Factory owners remained reliant on the accumulated experience and workplace co-operation of their machine-operators. How else can we explain the corporations' efforts to hold on to these workers and to reduce labour turnover? The Canadian steelmakers' experiments with welfare capitalism to stabilize their work force were part of a continent-wide movement to recast corporate management. Even Henry Ford introduced his celebrated "five-dollar day" to cut down the turnover in his auto plants.[12] Andrew Friedman has given the label "responsible autonomy" to the relative independence that managers gave to these skilled and semi-skilled production workers, but he is mistaken to see this primarily as a managerial strategy for controlling workers. For efficient production in any labour process requiring this level of skill and experience, the managers had no other choice.[13]

By extension, we should be quite careful not to inflate the impact of new management theories on the occupational structure. Undoubtedly in some industries the new bureaucratic approaches to centralizing control and subdividing labour were appealing to employers with particular labour problems. But it would be completely misleading to imagine scores of Canadian industrialists marching into their offices each morning clutching well-thumbed copies of *The Principles of Scientific Management*. F.W. Taylor's solutions could not be applied as widely as he insisted they could, and further research along the lines of this study will probably reveal that skilled workers and front-line supervisors in the heart of mass-production factories still retained discretionary power of the sort that he found reprehensible.[14]

A fourth conclusion reached in this study, which arises out of the third, has implications for an appreciation of working-class resistance and protest in the half century before World War II. Most accounts that emphasize the destruction of skill and working-class power have no adequate explanation for the rise of industrial unionism at the end of World War I and then again, more successfully, two decades later. Yet, the same kind of analysis that has been applied to the "crisis of the craftsman" in the late nineteenth and early twentieth centuries can help us understand the struggles of new categories of workers inside mass production. Specifically, we can identify workers whose skills and experience gave them a strategic advantage within the labour process, and who were able, in certain circumstances, to use their shop-floor autonomy and power and work-group solidarity to confront their employers with their own interests and concerns. Where they differed was in their lack of control over the labour market for their skills that the more independent craftsmen had enjoyed in the late nineteenth century. They soon learned that controlling the internal labour markets from which they were recruited required the unity of all workers in the plant – that is, industrial unionism. In automobile and electrical parts production, meatpacking, and pulp and paper work, for example, it was the more skilled men within the new production processes who organized and led the new unions in Canadian (and American) plants, as it would be again in the 1930s.[15] Working-class protest was far more likely to arise from these strata of the work force than from the poorest, most oppressed and downtrodden, and most defenceless. Paradoxically, though, the same kind of analysis can help us understand the frequent revolt of unskilled immigrant labourers in the early 1900s. These men could often use their leverage in a tight labour market to become some of the most troublesome workers in mass-production industry.

Fifth, if we move away from technologically determinist explanations of the momentous defeat of working-class activity in the early twentieth century, we are brought up against the same factors that were at work in the Canadian steel industry and the importance not simply of consciousness but of power. Specifically, the evidence presented here suggests that we should place at the centre of any explanation the patterns of vigorous repression in a context of labour surplus. Probably the most important change in this new world of work was the much greater clout of the corporate employers for imposing their will on their workers. In many industries, just as in steel, a rigid authoritarianism was implanted in which fear was meant to engender obedience.[16] This "spirit of Kaiserism," as some workers called it, and not the elaborate theorizing of industrial engineers like Taylor or the technological wizardry of "Fordism," was most important in undermining working-class resistance on the job. All the claims that have been made for Ford's technological solution to the "labour problem" have obscured the fact that the industrial espionage and security systems ("goon squads," in fact) in his plants were essential for maintaining discipline.[17] That kind of industrial tyranny was possible only when a rebellious worker could be dragged to the window to be reminded of the crowd of unemployed waiting at the gate to find work in the plant. The state collaborated in this repression by making available military and judicial support for keeping plants open in the case of mass picketing. It was only when the balance of power in the labour market shifted to the advantage of workers and the state felt compelled to adopt a new role of mediator and pacifier that workers were able to make any headway in the struggle for their collective interests. In the period under study, only the years of World War I and its immediate aftermath provided these conditions, and it would take a second world war to bring them back.

A sixth conclusion relates to the circumstances under which workers were able to break through the power of their corporate employers and assert their own concerns. Some edge in the labour market was clearly essential. Relatively full employment and rising wages could give workers some independence from their bosses. Yet those changes in material conditions were not necessarily enough. Some kind of ideological shift was needed to undermine the legitimacy of capitalist accumulation processes in workers' eyes and to establish new standards of justice and "fairness." The strident rhetoric of democracy and public service during World War I provided that new ideological climate and helped to fuel working-class resistance across the country. To create a new rupture in ideological hegemony on the same scale would take, first, the devastating economic slump of the 1930s and then the return of wartime rhetoric in the early 1940s.[18]

Finally, despite the impressive evidence of collective working-class resistance in this early twentieth-century period, we must recognize that the predominant response to new mass-production labour processes was personal, private, and informal. Down to at least 1920, many workers drifted in and out of these new plants, refusing to settle into jobs available to them. Employers' concern about labour turnover seemed to run through all manufacturing industries in these years, reaching a peak during World War I. These footloose habits made eastern workers look a lot like their western Canadian counterparts, and, indeed, itinerant workers flowed easily between regions in search of work. A great percentage of the Canadian working class in this period of rapid economic expansion (and contraction) were on the move – whether looking for adventure, purposefully seeking cash earnings to send home to agrarian families, or simply walking away in anger or disgust from unacceptable work experiences in many parts of the country.

Yet individualism could, and often did, take a different turn. Some workers were settling into their jobs and their communities, and by 1930 the turbulence of labour turnover had subsided considerably. Some writers have suggested this relative calm in industrial relations reflected a quiet contentment with the rewards of industrial capitalism. That conclusion, however, is too simple. In mass-production industries like steel, these workers' aspirations and expectations were shaped to a great extent by the hiring and promotion practices and welfare policies of their employers, and, after the colossal defeats of unions in the post-World War I period, by a fatalism about any alternative forms of industrial or social relations. All the repressive constraints of the workplace and the long-term welfare plans like pensions or life insurance could combine to encourage a quiet deference and "loyalism" to the corporation on which they were dependent. Whether through reluctant compliance or active sycophancy, workers found they had to make their own private accommodations with their bosses, especially their immediate supervisors, in order to assure a steady job, a regular income, and some long-term economic security. During the long spells when collective working-class action seemed impossibly fanciful or downright dangerous, Canadian factory workers accepted this "web of dependency." As an acceptance and internalization of the existing relations of social and economic power – what is often called "consent" – this was not a permanent state of mind for most, but it was a tenacious one for many. In some form or other, it was probably the predominant consciousness of factory workers in Canada between the wars.[19]

To point to the personal, private, and informal responses of mass-production workers is not to suggest a state of atomized, every-man-for-himself individualism. Most often, their accommodation to their subordination in capitalist industry fit into the plans of their families for collective survival in a society where wages were the only source of income. This study has not dealt extensively with the working-class family economy, but it can help to provide a deeper understanding of the constraints and possibilities provided by the world of work for the kinds of family strategies that other historians have recently begun to explore.[20] Of course, family life was more than a matter of economics. It was in their homes, and in their neighbourhoods generally, that many workers kept alive the alternative values of human dignity, mutual respect, and natural justice that their everyday job experience so often denied them.

The story of working-class response to mass-production industry did not end in the mid-1930s. Subsequent history would reveal just how much smouldering resentment could be buried in deferential behaviour. The remarkable success of industrial unionism in steel and other Canadian mass-production industries in the 1940s was based on the willingness of thousands of workers to act together, especially on the picket line, to re-assert the demands that had been raised at the end of World War I for "industrial democracy." Steelworkers took their first halting steps in the mid-1930s when independent unions appeared in Hamilton, Sault Ste. Marie, Sydney, and Trenton. These new initiatives flowed from a combination of experience gathered in the efforts to use industrial councils for working-class objectives and in organizing the unemployed in the steel towns in the early years of the Great Depression. Once again, the spark came from the more skilled workers, especially (though not exclusively) those immersed in Communist or social-democratic politics.[21]

In 1936 the Steel Workers Organizing Committee (SWOC) emerged within the new Committee on Industrial Organizations (CIO) in the United States and quickly won the affiliation of almost all the fledgling Canadian steelworkers' unions (the Algoma workers waited until 1940). In their rush to affiliate, these Canadian unionists were responding to the CIO's powerful appeal, but membership figures remained low in Canadian steel mills until the war. The independent Algoma union was closely tied to Sir James Dunn's revitalization plans for the Sault plant, and Sydney's local had only grudging and mean-spirited recognition from Disco. In a familar pattern, Stelco's tiny union was an insignificant presence in the plant. During the early years of the war, however, steelworkers' bargaining leverage shifted dramatically, and their resentments exploded in complete walkouts at Algoma and Disco in 1943. New collective bargaining legislation passed at the provincial and federal levels in 1943-1944 forced the steelmaking corporations to sign contracts with their workers' representatives. But a showdown was looming once wartime production was over and emergency legislation had expired. It occurred in the fall of 1946 in all three major steel centres, though the most critical battle was fought in Hamilton. The workers owed their victory at the end of more than eleven weeks to the dogged determination of a majority of steelworkers, especially the Europeans, and the active picket-line support of other workers in the city in keeping the plant gates closed. Henceforth, Canadian steelworkers would be able to have their union representatives negotiate important aspects of their terms of employment.[22]

The consolidation of industrial unionism in steel in the 1940s rested on several decisive factors that had not been present at the end of World War I. The absence of sojourners and the stability of the work force made waging such a battle much more practicable. This time the Europeans were solidly involved and committed to a unionism that promised them some relief from managerial tyranny. The union the steelworkers embraced was also a more effective instrument for an all-inclusive struggle than the old Amalgamated. An important participant in a buoyant industrial-unionist movement, SWOC and its successor, the United Steel Workers of America (USWA), were prepared to rally all workers in the industry regardless of skill or occupation. This was also a more centralized, bureaucratic organization, which undertook to co-ordinate bargaining strategies across the industry, particularly to introduce a common minimum wage for all steelworkers. After the war the union would similarly undertake to negotiate a radical simplification of wage rates through a "Co-operative Wage Study" with the steelmaking corporations.[23] Most important to the steelworkers' success in the 1940s, however, was the fact that the Canadian state stepped in decisively. The new legislation that required employers to recognize properly certified unions gave organized workers an enormous advantage that they had lacked a quarter of a century earlier.

The gains made in this new collective bargaining were generally those first articulated at the end of World War I – a reduction in hours (this time, a forty-hour week and paid vacations), new standards of fairness through seniority and grievance systems, and new procedures for negotiating some economic security. The administration of a Canadian steel plant was now regulated by bureaucratic rules set down in a collective agreement that would make the arbitrariness of the past much more difficult. Yet the earlier ambiguities in negotiating over subjects directly related to the labour process were swept away with new "management-rights" clauses. In effect, establishing a "rule of law" within industry did not mean any serious erosion of managerial power in the workplace; for the most part, it affected only the way that power was wielded. In the words of Stelco's historian, William Kilbourn, the corporation avoided an "abdication of a certain basic responsibility of management for determining work rules. . . . Sound seniority provisions were thus prevented from becoming mere rigid ritual codes for the consecration of inefficiency."[24]

While the steel companies and their managerial staff may have felt that their hands had been tied by the new collective agreements of the post-war era, they had kept for themselves the unilateral right to control the way steel production was organized and carried out. Consequently, if the 1940s were in one sense the culmination of a long series of struggles to entrench some "industrial democracy" in Canadian steel plants, they also marked the beginning of a new phase of work relations, in which the corporations would continue to look to technological and managerial innovations to increase productivity, relatively unhampered by union contracts. Steelworkers would never find these issues on the bargaining table, and the new bureaucratized structures of the Canadian industrial relations system made resistance to them more difficult and more likely to find expression in absenteeism, sabotage, or wildcat strikes.[25] In short, the struggle for "industrial democracy" would continue.

The steelworkers' efforts to meet the challenge of the Second Industrial Revolution still have resonance in the late twentieth century. In this age of the microchip, as a rising chorus of voices decries any resistance to the "Third" Industrial Revolution, we would do well to reflect on the lessons of that earlier factory regime in the age of mass production. The experience of those workers certainly suggests that major transformations in the world of work, especially dazzling technological changes, are no guarantee of happiness and prosperity for the workers involved. Without the power to assert their own needs and concerns, workers in those early twentieth-century factories had to face long hours, frequently dangerous working conditions, petty, autocratic supervision, and economic insecurity. Workers believed they had a right to expect better in return for their labour. And their struggles to make their workplaces more humane suggest that the path of innovation in the capitalist workplace is seldom smooth. The owners and managers of our modern "gigantic automatons" are doubtless in for some rough times.

Abbreviations

AISI	American Iron and Steel Institute
AJ	*Amalgamated Journal*
AUCAI	Algoma University College Archives, Interviews (names withheld)
BB	*Besco Bulletin*
BI	University College of Cape Breton, Beaton Institute
CBM	*Cape Breton's Magazine*
CBMR	Canada, Bureau of Mines, *Report on the Mining and Metallurgical Industries of Canada, 1907-8*
CE	*Canadian Engineer*
CF	*Canadian Foundryman*
CHR	*Canadian Historical Review*
CIMB	Canadian Institute of Mining and Metallurgy, *Bulletin*
CLL	*Canadian Labor Leader* (Sydney)
CM	*Canadian Machinery*
CMJ	*Canadian Mining Journal*
CMM	*Canadian Mining Manual*
CMR	*Canadian Mining Review*
DUA	Dalhousie University Archives
EF	*Eastern Federationist* (New Glasgow)
IC	*Industrial Canada*
ISC	*Iron and Steel of Canada*
IWMR	Interchurch World Movement, Commission of Inquiry, *Report on the Steel Strike of 1919* (New York, 1920)
LG	*Labour Gazette*
L/LT	*Labour/Le Travail* (formerly *Labour/Le Travailleur*)
LN	*Labor News* (Hamilton)
MLH	*Maritime Labor Herald*
ND	*New Democracy* (Hamilton)
NSCHLR	Nova Scotia, Commission on Hours of Labor, *Report* (1910)
NSDMR	Nova Scotia, Department of Mines, *Report*
OBMR	Ontario, Bureau of Mines, *Report*
OIFR	Ontario, Inspectors of Factories, *Report*
PAC	Public Archives of Canada
PANS	Public Archives of Nova Scotia
PAO	Provincial Archives of Ontario
RCIRE	Canada, Royal Commission of Industrial Relations, "Evidence"
RCRCL	Canada, Royal Commission on the Relations of Capital and Labour
SF	*Stelco Flashes*
WW	*Workers' Weekly* (Stellarton)

Notes

Introduction

1. This term won formal recognition as a shorthand for a new form of manufacturing in 1926, when the Encyclopaedia Britannica asked Henry Ford to contribute an entry on "mass production" to its three-volume supplement. The article was ghost-written by a Ford Company spokesperson, William J. Cameron, and appeared over Ford's name. See David A. Hounshell, *From the American System to Mass Production, 1800-1932: The Development of Manufacturing Technology in the United States* (Baltimore, 1984), p. 1.

2. Terry Copp, *The Anatomy of Poverty: The Condition of the Working Class in Montreal, 1897-1929* (Toronto, 1974); Michael J. Piva, *The Condition of the Working Class in Toronto – 1900-1921* (Ottawa, 1979); Paul Larocque, "Aperçu de la condition ouvrière à Québec, 1896-1914," *L/LT*, 1 (1976), pp. 122-38; Eleanor A. Bartlett, "Real Wages and the Standard of Living in Vancouver, 1901-1929," *B.C. Studies*, 51 (Autumn, 1981), pp. 3-62.

3. For a fuller discussion of the periodization of industrial development in Canada, see Craig Heron and Robert Storey, "On the Job in Canada," in Heron and Storey, eds., *On the Job: Confronting the Labour Process in Canada* (Kingston and Montreal, 1986), pp. 5-26; and Gregory S. Kealey, "The Structure of Canadian Working-Class History," in W.J.C. Cherwinski and G.S. Kealey, eds., *Lectures in Canadian Labour and Working-Class History* (St. John's, 1985), pp. 23-36.

4. Some of the more important studies include W.J.A. Donald, *The Canadian Iron and Steel Industry: A Study in the Economic History of a Protected Industry* (Boston, 1915); Edward J. McCracken, "The Steel Industry of Nova Scotia" (M.A. thesis, McGill University, 1932); Donald Eldon, "American Influence in the Canadian Iron and Steel Industry" (Ph.D. thesis, Harvard University, 1952); Krys Inwood, *The Canadian Charcoal Iron Industry, 1870-1914* (New York, 1986); James M. Cameron, *Industrial History of the New Glasgow District* (New Glasgow, n.d.); William Kilbourn, *The Elements Combined: A History of the Steel Company of Canada* (Toronto, 1960); Duncan McDowall, *Steel at the Sault: Francis H. Clergue, Sir James Dunn, and the Algoma Steel Corporation, 1901-1956* (Toronto, 1984); Ronald F. Crawley, "Class Conflict and the Establishment of the Sydney Steel Industry, 1899-1904" (M.A. thesis, Dalhousie University, 1980); Arthur Martin Kruger, "Labour Organization and Collective Bargaining in the Canadian Basic Steel Industry" (Ph.D. thesis, Massachusetts Institute of Technology, 1959); Robert McDonald Adams, "The Development of the United Steelworkers in Canada, 1936-1951" (M.A. thesis, Queen's University, 1952); Robert Henry Storey, "Workers, Unions, and Steel: The

Reshaping of the Hamilton Working Class, 1935-1948" (Ph.D. thesis, University of Toronto, 1981).

5. These formed the basis of David Brody's exemplary study of work and workers, *Steelworkers in America: The Nonunion Era* (New York, 1960).

Chapter 1

1. A recent comparison of the British and American industries has highlighted just how different the industry could be between two world leaders; see Bernard Elbaum and Frank Wilkinson, "Industrial Relations and Uneven Development: A Comparative Study of the American and British Steel Industries," *Cambridge Journal of Economics*, 3 (1979), pp. 275-303.

2. *CMR*, 16,2 (February, 1897), p. 51.

3. The development of these early enterprises is discussed in H. Clare Pentland, *Labour and Capital in Canada, 1650-1860* (Toronto, 1981), pp. 34-46; Louise Tessier, *Les Forges du Saint-Maurice: Their Historiography* (Ottawa, 1980); James Herbert Bartlett, *The Manufacture, Consumption and Production of Iron, Steel, and Coal in the Dominion of Canada, With Some Notes on the Manufacture of Iron, and on the Iron Trade, in Other Countries* (Montreal, 1885); Donald, *Canadian Iron and Steel Industry*; Edward J. McCracken, "The Steel Industry of Nova Scotia" (M.A. thesis, McGill University,1932), pp. 39-65; Christopher Alfred Andreae, "A History of Nineteenth Century Iron Works in Nova Scotia" (M.Museum Studies thesis, University of Toronto, 1981); Dianne Newell, *Technology on the Frontier: Mining in Old Ontario* (Vancouver, 1986), pp. 93-103.

4. Inwood, *Canadian Charcoal Iron Industry*.

5. Pentland, *Labour and Capital*, pp. 42-46; Robert Gourlay, *Statistical Account of Upper Canada* (2 vols., London, 1822), I, pp. 324-26; Elijah Leonard, *A Memoir* (London, Ont., n.d.), p. 6; Edward Allen Talbot, *Five Years Residence in the Canadas*; Steel Company of Canada, Ltd., *Reports Upon the Property at Londonderry, Nova Scotia, with Analyses of Ores, &c.* (London, 1874), pp. 53, 57; Andreae, "Iron Works," pp. 11-14, 21, 23, 85; David E. Stephens, "Boom-town of Iron and Steel," *Nova Scotia Historical Quarterly*, 4,1 (March, 1974), pp. 24-25; Donald, *Canadian Iron and Steel Industry*, pp. 72-73.

6. *CMM*, 1897, p. 79; PAC, RG 87, Vol. 18, File 81, Statements of Dartmouth and London rolling mills (1910) and Manitoba Rolling Mill Company (1911).

7. Donald, *Canadian Iron and Steel Industry*; McDowall, *Steel at the Sault*, pp. 8-22; Tom Traves, *The State and Enterprise: Canadian Manufacturers and the Federal Government, 1917-1931* (Toronto, 1979), pp. 121-54.

8. DUA, MG 4, 106 (Hawker-Siddeley Papers), Deeds, Agreements, Charters, etc., 1882-1920, File: Charters, Patents, "Prospectus of the Nova Scotia Steel Company"; PANS, MG 1, Vol. 2155 (J. Dix Fraser, "Graham Fraser – His Life and Work"); MG 3, 1877 (Dominion Steel and Coal Papers), 11a (Nova Scotia Steel and Coal Company Limited, *Annual Reports for Years 1883 to 1900 (inclusive)* [New Glasgow, n.d.]); Cameron, *New Glasgow*, pp. III1-III4; L.D. McCann, "The Mercantile-Industrial Transition in the Metal Towns of Pictou County, 1857-1931," *Acadiensis*, 10,2 (Spring, 1981), pp. 40-50; McCracken,

"Steel Industry," pp. 81-83; Andreae, "Ironworks," pp. 131-45; L. Anders Sandberg, "The Deindustrialization of Pictou County, Nova Scotia – Capital, Labour, and the Process of Regional Decline, 1881-1921" (Ph.D. thesis, McGill University, 1985).

9. PANS, MG 3, 1877, Nova Scotia Steel and Coal Company Limited, *Annual Reports*; C.W. Vernon, *Cape Breton, Canada, at the Beginning of the Twentieth Century: A Treatise of Natural Resources and Development* (Toronto, 1902), pp. 178-84; McCann, "Mercantile-Industrial Transition," pp. 50-57; Sandberg, "Deindustrialization"; Donald, *Canadian Iron and Steel Industry*, pp. 195-99; M.O. Hammond, "The Industrial Development of Nova Scotia," *Globe*, 21 April 1913; James Frost, "The 'Nationalization' of the Bank of Nova Scotia, 1880-1910," *Acadiensis*, 12,1 (Autumn, 1982), p. 25; David Frank, "The Cape Breton Coal Industry and the Rise and Fall of the British Empire Steel Corporation," *ibid.*, 7,1 (Autumn, 1977), pp. 14-15.

10. David Carnegie, *The History of Munitions Supply in Canada, 1914-1918* (London, 1918).

11. *Canadian Gazette*, 8 February 1913; Donald, *Canadian Iron and Steel Industry*, p. 199.

12. Hamilton *Spectator*, 31 December 1895; W.A. Child, "Iron Trade Built by Determined Men," *ibid.*, 15 July 1926; Kilbourn, *History of the Steel Company of Canada*, pp. 51-112; W. Craig Heron, "Working-Class Hamilton, 1895-1930" (Ph.D. thesis, Dalhousie University, 1981), pp. 2, 14-15, 19.

13. *CMR*, 18,10 (October, 1899), p. 257; 18,6 (June, 1899), p. 189; Watson Griffin, *At the Front Door of Canada: The Great Works of the Dominion Iron and Steel Company at Sydney, C.B.* (Montreal, 1899), pp. 53-58; *CE*, 8,18 (October, 1901), p. 433; A.J. Moxham, *Canada as a Steel Producer: The Operations of the Dominion Iron & Steel Company* (n.p., n.d.), p. 13; James Stephen Jeans, *Canada's Resources and Possibilities, With Special Reference to the Iron and Allied Industries, and the Increase of Trade with the Mother Country* (London, 1904), pp. 123, 131; Don Macgillivray, "Henry Melville Whitney Comes to Cape Breton: The Saga of a Gilded Age Entrepreneur," *Acadiensis*, 9,1 (Autumn, 1979), pp. 64-67.

14. *CMR*, 18,11 (November, 1899), p. 289; Griffin, *Front Door*, p. 53; *IC*, 2,9 (May, 1902), p. 322; Vernon, *Cape Breton*, pp. 206-21.

15. *CMR*, 21,3 (March, 1902), pp. 45-46; 22,4 (April, 1903), pp. 77-78; 22,5 (May, 1903), pp. 101-02; 22,12 (December, 1903), p. 241; 23,10 (October, 1904); Jeans, *Canada's Resources*, pp. 123-31.

16. *LG*, 4,3 (September, 1903), p. 185; 6,1 (July, 1905), p. 16; Donald, *Canadian Iron and Steel Industry*, pp. 200-11; *ISC*, 2,5 (June, 1919), pp. 144-45.

17. *CE*, 8,10 (February, 1901), pp. 210-12; McDowall, *Steel at the Sault*, pp. 23-49; Sault Ste. Marie Museum, Archives, 240.5, MS (Donald Machum, "History of Algoma Steel and Related Companies"), pp. 8-108; Donald Eldon, "The Career of Francis H. Clergue," *Explorations in Entrepreneurial History*, 3,4 (15 April 1951), pp. 254-62; Margaret Van Emery, "Francis Hector Clergue and the Rise of Sault Ste. Marie as an Industrial Centre," *Ontario History*, 56,3 (September, 1964), pp. 191-98; John Ferris, *Algoma's Industrial and Trade Union Development* (Sault Ste. Marie, n.d.), pp. 11-23; H.V. Nelles, *The Politics of Develop-*

ment: Forests, Mines & Hydro-Electric Power in Ontario, 1894-1941 (Toronto, 1974), pp. 126-34; Alan Sullivan, *The Rapids* (Toronto, 1972 [1922]).

18. Walter Isard, "Some Locational Factors in the Iron and Steel Industry Since the Early Nineteenth Century," *Journal of Political Economy* (1948), pp. 203-04; OBMR, 1901, pp. 63-64; 1908, p. 297; *IC*, 2,9 (May, 1902), pp. 325-27; *CMM*, (1902), pp. 425-27; (1903), pp. 88-90; PAC, RG 36/17, Vol. 5, p. 858; McDowall, *Steel at the Sault*; Machum, "History," pp. 69-71; Eldon, "Clergue," p. 263. Clergue's blunder in choosing the Bessemer process is highlighted by the fact that only one new Bessemer plant was erected in the United States in the decade before 1905. Victor S. Clark, *History of Manufactures in the United States, Volume III: 1893-1928* (New York, 1929), p. 70

19. McDowall, *Steel at the Sault*; Machum, "History," pp. 78-105; Van Emery, "Clergue," pp. 198-201; Eldon, "Clergue," pp. 262-64; Ferris, *Algoma's Industrial and Trade Union Development*, pp. 28-35, 104-05; Nelles, *Politics of Development*, pp. 134-38; Livo Ducin, "Labour's Emergent Years and the 1903 Riots," in *50 Years of Labour in Algoma: Essays on Aspects of Algoma's Working-Class History* (Sault Ste. Marie, 1978), pp. 1-14.

20. As Duncan McDowall has explained, the English group was, sensibly, interested in turning the corporation into a diversified, integrated steel works, while the Philadelphia interests were anxious to limit new capital investments in order to skim off more profits. The English were in control from 1909 to 1916.

21. McDowall, *Steel at the Sault*, pp. 50-68; Machum, "History," pp. 109-213; Ferris, *Algoma's Industrial and Trade Union Development*, pp. 35-80.

22. *CMM* (1899), p. 175; *CE*, 6,10 (February, 1899), pp. 284-85; 8,10 (February, 1901), pp. 204-05; 16,13 (26 March 1909), pp. 425-28; OBMR, 1902, pp. 23-24; 1908, pp. 314-24; *IC*, 2,9 (May, 1902), pp. 331-34; Jeans, *Canada's Resources*, pp. 147-48; CBMR, pp. 321, 332-35; *CMJ*, 2,5 (1 March 1908), pp. 36-39; *ISC*, 4,4 (May, 1921), p. 109; Donald, *Canadian Iron and Steel Industry*, pp. 228-32, 237; Eldon, "American Influence," pp. 96-97, 131; F.W. Gray, "Iron Smelting and Steel Making in Canada: A Depressing and Disastrous Record," *ISC*, 2,12 (January, 1920), p. 4.

23. *IC*, 2,9 (May, 1902), pp. 329-30; *CMM*, 1900, pp. 65-67; OBMR, 1901, pp. 27-30; 1908, pp. 308-13; CBMR, p. 322; W.R. Williams, "The Midland Blast Furnace," *Inland Seas*, 19,4 (Winter, 1963), pp. 308-10; Eldon, "American Influence," p. 97; Donald, *Canadian Iron and Steel Industry*, pp. 222-27; Halifax *Herald*, 22-25 October, 24-25 November 1902; *CMR*, 22,6 (June, 1903), p. 124; 23,2 (February, 1904), p. 32; CBMR, pp. 539-43; Archibald McColl, "Nova Scotia's Industrial Development," *Evening News* (New Glasgow), 18 December 1914; McCracken, "Steel Industry," p. 62; *LG*, 9,7 (January, 1909), p. 757; PANS, MG1, no. 170 (Thomas Cantley Papers, Letterbook), p. 421 (T. Cantley to H.A. Wilson, 10 November 1908); *CMM*, 10,9 (28 August 1913), p. 219.

24. OBMR, 1908, pp. 325-26; Donald, *Canadian Iron and Steel Industry*, pp. 227-28, 232-34; Eldon, "American Influence," pp. 98-99; *CM*, 7,11 (November, 1911), pp. 293-95; 8,11 (November, 1912), pp. 325-30; 9,1 (2 January 1913), pp. 20-24; 10,26 (25 December 1913), pp. 651-57; *ISC*, 3,1 (February, 1920), p. 30; John Bradbury, unpublished manuscript, pp. 312-13.

25. Canada, Dominion Bureau of Statistics, *Iron and Steel and Their Products* (Ottawa, 1930), pp. 68, 73.

26. *ISC*, 2,8 (September, 1919), pp. 199-200.

27. McDowall, *Steel at the Sault*, pp. 69-94; Machum, "History," pp. 193-223; *Amplified Brief of the Algoma Steel Corporation, Limited* (n.d. [1928], n.p.); Traves, *State and Enterprise*, pp. 121-54.

28. Frank, "British Empire Steel Corporation," pp. 3-34; McCracken, "Steel Industry," pp. 180-232; *ISC*, 2,7 (August, 1919), p. 193; (September, 1919), p. 217; (March, 1920), p. 72; (June, 1920), p. 155; Ernest R. Forbes, "Misguided Symmetry: The Destruction of Regional Transportation Policy for the Maritimes," in David Jay Bercuson, ed., *Canada and the Burden of Unity* (Toronto, 1977), pp. 60-86.

29. McCann, "Mercantile-Industrial Transition," pp. 57-64; DUA, MG 4, 106, General Correspondence, 1919-25, File: Canadian Car and Foundry, letter to D.H. McDougall, 17 June 1921; PANS, MG 2, Vol. 11B (E.A. Armstrong Papers), F5/4093 (George White and A.T. Logan to Sir Henry Thornton, 29 September 1924); Halifax *Herald*, 26 March, 29 April 1921; *IC*, 20,9 (January, 1920), p. 177; 21,3 (June, 1920), p. 65; RCIRE, New Glasgow, p. 3581; Nova Scotia, Royal Commission on Provincial Development and Rehabilitation, *Report on the Manufacturing of Secondary Iron and Steel Products and its Relation to the Primary Iron and Steel Industry in the Province of Nova Scotia* (Halifax, 1944); Nova Scotia, Commissioner on Trenton Steel Works, *Report* (Halifax, 1944).

30. Arthur G. McKee & Company, "Report on the Manufacture of Iron and Steel Products and Its Relation to the Primary Iron and Steel Industry in the Province of Nova Scotia, Canada," in Nova Scotia, Royal Commission on Provincial Development and Rehabilitation, *Report* (2 vols., Halifax, 1944), II, p. 48; Traves, *State and Enterprise*, pp. 122-39; Donald P. Kerr, "The Location of the Iron and Steel Industry in Canada," in R. Louis Gentilcore, ed., *Geographical Approaches to Canadian Problems* (Scarborough, 1971), pp. 59-67.

31. Kilbourn, *History of the Steel Company of Canada*, pp. 105, 111-12, 117-19, 126-27, 140; Steel Company of Canada, *Annual Reports* (Hamilton), 1902-22.

32. Donald, *Canadian Iron and Steel Industry*.

33. *Ibid.*, p. 139; Glen Williams, *Not For Export: Toward a Political Economy of Canada's Arrested Industrialization* (Toronto, 1983).

34. PAC, RG 36/11, Vol. 2, File 2-8; Traves, *State and Enterprise*, pp. 121-54; McDowall, *Steel at the Sault*, pp. 69-94, 124-77; McKee & Company, "Report."

35. The British business journalist James Stephen Jeans, who visited Canadian steel plants at the turn of the century, noted, "Canada is yet too 'young' to have a class of trained managers such as there is in England and in the United States. There is no class of managers that can be called typically Canadian, as there is one distinct class for England and another known as American. The Canadian managers of to-day are brought from many different lands and localities. Perhaps, on the whole, the greater number of them are imported from the United States. This remark certainly applies to the managers of coal mines, iron ore mines, and iron and steel works." Jeans, *Canada's Resources*, p. 206.

36. *CMM*, 1928.

37. Canada, House of Commons, *Journals*, 45 (1909-10), Appendix, Part III ("Proceedings of the Special Committee on Bill No. 21, 'An Act Respecting Hours of Labour on Public Works,' Comprising Reports, Evidence and Correspondence, December 9, 1909-May 1910"); *Complete Evidence and Statements of Conciliation Court between Workmen and Management of the Nova Scotia Steel & Coal Company and Eastern Car Company* (hereafter *Conciliation Court*) (n.p., n.d. [New Glasgow, 1915]).

38. See, for example, *Amplified Brief*, pp. 14-16. In 1928 the average hourly wage for skilled labour in Canadian steel plants was calculated to be 59.8 cents, compared to 74.75 in the United States. PAC, RG 36, 11, Vol. 1, File 2-3, "Iron and Steel Wages: Comparisons – 1928, Europe and America."

39. *LG*, 25,1 (January, 1928), p. 63; *Amplified Brief*, p. 7; PAC, RG 36/11, Vol. 8, File 3-6, "Report to Dominion of Canada Advisory Tarriff [*sic*] Commission Board as Concerning Employes of Algoma Steel Corporation, Limited, Sault Ste. Marie, Ontario"; author's interview with Earl Orchard, Sault Ste. Marie, 12 July 1984; AUCAI; Sydney *Post*, 23 March 1929; George MacEachern, "Autobiography" (typescript), p. 27

40. McCann, "Mercantile-Industrial Transition"; Cameron, *New Glasgow*; Ralph Murray Guy, "Industrial Development and Urbanization of Pictou County to 1900" (M.A. thesis, Acadia University, 1962).

41. Vernon, *Cape Breton*; Charles W. Dunn, *Highland Settler: A Portrait of the Scottish Gael in Nova Scotia* (Toronto, 1953); Del Muise, "The Making of an Industrial Community: Cape Breton Coal Towns, 1867-1900," in Don Macgillivray and Brian Tennyson, eds., *Cape Breton Historical Essays* (Sydney, 1980), pp.76-94; David Frank, "Tradition and Culture in the Cape Breton Mining Community in the Early Twentieth Century," in Kenneth Donovan, ed., *Cape Breton at 200: Historical Essays in Honour of the Island's Bicentennial, 1785-1985* (Sydney, 1985), pp. 203-18; David Frank and Donald Macgillivray, "Introduction," in Dawn Fraser, *Echoes From Labor's War: Industrial Cape Breton in the 1920s* (Toronto, 1976), pp. 9-23.

42. "Sault Ste. Marie and Its Industries," *CE*, 10,1 (January, 1903), pp. 15-20; L.L. Prior, "Sault Ste. Marie and the Algoma Steel Corporation Ltd." (M.A. thesis, University of Toronto, 1956); McDowall, *Steel at the Sault*; Van Emery, "Clergue."

43. Heron, "Working-Class Hamilton," pp. 1-76; Kilbourn, *History of the Steel Company of Canada*.

Chapter 2

1. There is now an extensive antiquarian and modern literature on the early iron works in Canada. Some of the most useful in the following summary of pre-industrial ironmaking are: Benjamin Sulte, *Les Forges Saint-Maurice* (Montreal, 1920); Elijah Leonard, *A Memoir* (London, n.d.), p. 6; W.J. Patterson, "The Long Point Furnace," Ontario Historical Society, *Papers and Records*, 36 (1944), p. 73; Andreae, "Iron Works"; B.J. Harrington, "Notes on the Iron Ores of Canada and Their Development," in Canada, Geological Survey, *Report of*

Progress (Montreal, 1873-74), p. 253; *CE* 1,1 (March, 1894), p. 310; 2,10 (February, 1895), pp. 288-89; 2,11 (March, 1895), pp. 310-12; *CMR*, 15,1 (January, 1896), p. 11; 15,8 (August, 1896), p. 172; 16,2 (February, 1897), p. 47; 18,3 (March, 1899), p. 57; CBMR; André Bérubé, "Technological Changes at Les Forges du Saint-Maurice, Quebec, 1729-1883," CIMB, 76,863 (May, 1983), p. 3; R.R. Potter, "The Woodstock Iron Works, Carleton County, New Brunswick," *ibid.*, p. 10; David J. McDougall, "The Grantham Iron Works," *ibid.*, 76,855 (July, 1983), p. 55; A.R.R. Jones, "A Gigantic Automaton," *ISC*, 7,4 (April, 1924), p. 64; Peter Temin, *Iron and Steel in Nineteenth-Century America: An Economic Inquiry* (Cambridge, 1964).

2. Hamilton *Spectator*, 10 February 1896.

3. Bérubé, "Technological Changes," p. 3; Eric Arthur and Thomas Ritchie, *Iron: Cast and Wrought Iron in Canada from the Seventeenth Century to the Present* (Toronto, 1982), pp. 1-3; Sulte, *Forges*, p. 100; F.C. Wertele, "Historical Record of the Saint Maurice Forges, the Oldest Active Blast Furnace on the Continent of America," Royal Society of Canada, *Proceedings and Transactions*, 4 (1886), Section II, pp. 80-81; T. Ritchie, "Joseph Van Norman, Ironmaster of Upper Canada," *Canadian Geographical Journal*, 77,2 (August, 1968), p. 48; Harrington, "Notes," pp. 242-59; Henry How, *The Mineralogy of Nova Scotia: A Report to the Provincial Government* (Halifax, 1869), pp. 84-85; NSDMR, 1877, pp. 45-46; Michel Gaumond, *Les Forges de Saint-Maurice* (Quebec, 1969), pp. 11-13; Dube, *Les vielles forges*, pp. 21-51; J.A. Bannister, "The Houghton Iron Works," Ontario Historical Society, *Papers and Records*, 26 (1944), pp. 80-81; Patterson, "Long Point Furnace," pp. 73-76; Owen, *Pioneer Sketches*, 452-57; Leonard, *Memoir*; R.W. Ellis, "Iron Deposits of Carleton County, New Brunswick," in Canada, Geological Survey, *Report of Progress*, 1874-75, p. 103; Potter, "Woodstock Iron Works," p. 10; David J. McDougall, "The Grantham Iron Works," *ibid.*, 76,855 (July, 1983), p. 55; Andreae, "Iron Works," p. 14; John Birkinbine, "The Old and New Iron Industry Compared," *CMR*, 21,4 (April, 1902), p. 90; Jones, "Gigantic Automaton," p. 64; Hamilton *Spectator*, 10 February 1896.

4. Bérubé, "Technological Changes," p. 3; Arthur and Ritchie, *Iron*, p. 3; Temin, *Iron and Steel in Nineteenth-Century America*, pp. 16-17.

5. Harrington, "Notes," pp. 247, 249-50, 252; Bérubé, "Technological Changes," p. 8; Potter, "Woodstock Iron Works," p. 10; D.D. Hogarth, "The Hull Iron Range, 1801-1977," *CIMB*, 76,854 (June, 1983), p. 22; McDougall, "Grantham Iron Works," p. 55; *CMM*, 1897.

6. *CMR*, 15,2 (February, 1896), p. 39; *CE*, 3,7 (January, 1896), pp. 248-49; Hamilton *Spectator*, 10 February 1896; Charles Reitell, *Machinery and its Benefits to Labor in the Crude Iron and Steel Industries* (Menasha, Wisc., 1917), pp. 9-21; OIFR, 1896, p. 11; *SF*, 14,6 (June, 1950). The task of carrying the iron pigs into railway cars provided F.W. Taylor, the celebrated theorist of scientific management, with one of his most famous experiments, involving Schmidt, the dull-witted Dutch labourer; see F.W. Taylor, *The Principles of Scientific Management* (New York, 1967), pp. 41-47.

7. Toronto *Leader*, 8 May 1860; Montreal *Transcript*, 7 March 1861; *Reports and Testimentials Relating to the Iron Mines of Nova Scotia* (Halifax, 1855), p. 30

(copy in PANS, MG 1, Vol. 260, p. 4); PANS, MG 100, Vol. 177, 16 (E.A. Jones to J.P. Edwards, 17 April 1914), p. 4; Andreae, "Iron Works," pp. 29-73; *CMM*, 1897, pp. 74, 79, 110; *Iron Age* (New York), 4 July 1895, 30 April 1899, 5 April 1900; Canada, Department of Trade and Commerce, *Report* (Ottawa), 1908, Part 1, pp. 796-97; *SF*, 14,6 (June, 1950), p. 19.

8. James J. Davis, *The Iron Puddler: My Life in the Rolling Mills and What Came of It* (Indianapolis, 1922); John A. Fitch, *The Steel Workers* (New York, 1910), pp. 32-37; *CE*, 22 (4 January 1912), p. 124; *CM*, 22,19 (6 November 1919), p. 462; *SF*, 14,6 (June, 1950), p. 19.

9. *CMM*, 1890-91, p. 149; see also Clark, *History of Manufactures*, p. 82.

10. Jones, "Gigantic Automaton," p. 100; Cameron, *New Glasgow*, p. V2. See also Patrick McGeown, *Heat the Furnace Seven Times More* (London, 1967), pp. 79-80; Reitell, *Machinery*, pp. 23-24.

11. Stephens, "Boomtown," p. 26; Krys Inwood, "Inter-sectoral Linkages: the Case of Secondary Iron and Steel" (paper presented to the Business History Conference, Trent University, May, 1984); Donald, *Canadian Iron and Steel Industry*.

12. For a discussion of Armstrong's painting, see Gregory S. Kealey, "Toronto Rolling Mills," Committee on Canadian Labour History, *Newsletter*, 7 (1975), pp. 1-2.

13. Cameron, *New Glasgow*, pp. V15-V16; RCRCL, *Evidence – Nova Scotia* (Ottawa, 1889), pp. 258, 266; *Morning Herald* (Halifax), 30 November 1881; W.A. Child, "Iron Trade Built By Determined Men," Hamilton *Spectator*, 15 July 1926.

14. Jones, "Gigantic Automaton," pp. 62-63; W.S. Wilson and A.P. Theurkauf, "What the Eyes Behold at Sydney Steel Plant," *ISC*, 10, 11 (November, 1927), p. 335; Ralph D. Williams, *The Honorable Peter White: A Biographical Sketch of the Lake Superior Iron Country* (Cleveland, 1905), pp. 146-60, 181; Reitell, *Machinery*, pp. 9-10; *CE*, 5,2 (August, 1897), p. 120; *SF*, 14,6 (June, 1950), p. 7; Gillies-Guy Company Archives (Hamilton), H.W. Robinson Typescript, 14 November 1952; E.W. Hanna, "Steel Making in Cape Breton: A Comprehensive Description of Eastern Canada's Greatest Industry," *Cape Breton Magazine*, 1,2-3 (October-November, 1901), p. 56; Vernon, *Cape Breton*, p. 217; *CMM*, 1901, pp. 129-30; *IC*, 2,9 (May, 1902), p. 323; *CMR*, 21,7 (July, 1902), p. 195; Sydney *Post*, 30 August 1904; OBMR, 1908, p. 291; CBMR, pp. 325, 528, 559; *CMJ*, 35,13 (15 July 1914), p. 487; *ISC*, 2,10 (November, 1919), p. 262; 10,9 (September, 1927), p. 284; *Besco Bulletin*, 73 (26 June 1926), p. 3.

15. NSDMR, 1893, p. 40; *CMM*, 1901, pp. 130, 133-34; Moxham, *Canada as a Steel Producer*, p. 39; Hanna, "Steel Making in Cape Breton," p. 59; Vernon, *Cape Breton*, pp. 182, 217; Sydney *Post*, 30 August 1904; CBMR, pp. 534-35; *CMJ*, 32, 11 (1 June 1911), pp. 333-40; 33, 18 (15 September 1912), pp. 641-43; 35,13 (15 July 1914), p. 487; *IC*, 2,9 (May, 1902), pp. 324, 326; *ISC*, 2,10 (November, 1919), pp. 262-68; 2,11 (December, 1919), pp. 291-97; Jones, "Gigantic Automaton," pp. 62-64; Wilson and Theurkauf, "What the Eyes Behold," p. 334.

16. *CMM*, 1901, pp. 130-31; *CMR*, 20,10 (October, 1901), pp. 240-46; 21,12 (December, 1902), pp. 291-300; Moxham, *Canada as a Steel Producer*, p. 8; Hanna, "Steel Making in Cape Breton," p. 57; OBMR, 1908, pp. 291-97, 301-04; CBMR, pp. 337, 528-30, 546-47; Stelco, *Annual Report*, 1920, n.p.; *CF*,

19,2 (February, 1928), pp. 7-10; Jones, "Gigantic Automaton," pp. 63-64; Wilson and Theurkauf, "What the Eyes Behold," pp. 335-36; Charles Rumford Walker, *Steel: The Diary of a Furnace Worker* (Boston, 1922), pp. 81-113.

17. David S. Landes, *The Unbound Prometheus: Technological Change and Industrial Development in Western Europe from 1750 to the Present* (Cambridge, 1972), p. 255; Fitch, *Steel Workers*, pp. 39-42.

18. *Sault Star*, 20 February 1902. See also OBMR, 1901, p. 64.

19. Algoma finally abandoned its Bessemer converters in 1916. Disco installed two of them in 1907 to be used in a hybrid process known as "duplex" steelmaking, wherein pig iron was first heated in a Bessemer converter and then transferred to an open-hearth furnace for a short time to complete the refining. Although more expensive, this procedure took much less time than normal open-hearth production and allowed for a speed-up of steel production when demand was heavy. See A.P. Scott, "The Duplex Process at Sydney, N.S.," *CMJ*, 33,18 (15 September 1912), p. 632; *ISC*, January, 1926, p. 11; Clark, *History of Manufactures*, III, p. 78.

20. Jones, "Gigantic Automaton," pp. 100-03; Wilson and Theurkauf, "What the Eyes Behold," pp. 336-37; Reitell, *Machinery*, pp. 21-29; Frank Popplewell, *Some Modern Conditions and Recent Developments in Iron and Steel Production in America* (Manchester, 1906), p. 96; John Fitch, *The Steel Workers* (New York, 1910), p. 46; *Iron Age*, 5 July 1900, p. 7; *CE*, 8,1 (July, 1900), pp. 46-47; 22 (15 February 1912), pp. 306-07; Hanna, "Steel Making in Cape Breton," pp. 60-61; *CMM*, 1901, pp. 132-33; OBMR, 1908, pp. 297-99; CBMR, pp. 530-31, 548-49; *CMJ*, 34,5 (1 August 1913), pp. 488-89; *CM*, 9,1 (2 January 1913), p. 56; *CF*, 6,12 (December, 1915), p. 217; Stelco, *Annual Reports*, 1912-1913, n.p.; Walker, *Steel*, pp. 16-80.

21. Jones, "Gigantic Automaton," p. 118; Wilson and Theurkauf, "What the Eyes Behold," pp. 337-40; Hanna, "Steel Making in Cape Breton," p. 62; Jeans, *Canada's Resources and Possibilities*, p. 141; *IC*, 2,9 (May, 1902), p. 325; *CE*, 13,2 (February, 1906), pp. 38-43; 14 (31 July 1913), pp. 251-52; OBMR, 1908, p. 299; CBMR, pp. 326, 532-33; *CF*, 4,9 (September, 1913), pp. 142-44; *CMJ*, 24,5 (1 August 1913), p. 489; *ISC*, January, 1920, pp. 324-27; May, 1920, pp. 113-19; Fitch, *Steel Workers*, pp. 45-56.

22. Canada, House of Commons, *Journals*, Vol. 45 (1909-10), Appendix, Part III ("Proceedings of a Special Committee on Bill No. 21, 'An Act Respecting Hours of Labour on Public Works,' Comprising Reports, Evidence and Correspondence, December 9, 1909 – May 3, 1910") (hereafter "Eight-Hour Committee"), p. 183; *Conciliation Court*; RCIRE, New Glasgow, pp. 3606-07; AUCAI; *SF*, 14,6 (June, 1950), p. 9; Fitch, *Steel Workers*, pp. 49-53; Whiting Williams, *What's On the Worker's Mind, By One Who Put on Overalls to Find Out* (New York, 1920), pp. 41-70.

23. For the more recent changes of the "Third Industrial Revolution," see Sin Tze Ker, "Technological Change in the Canadian Iron and Steel Mills Industry, 1946-69" (Ph.D. thesis, University of Manitoba, 1972); and Ontario Task Force on Employment and New Technology, *Employment and New Technology, Appendix 5: Employment and New Technology in the Iron and Steel Industry* (Toronto, 1985).

24. Arthur G. McKee & Company, "Report on the Manufacture of Iron and Steel Products and its Relation to the Primary Iron and Steel Industry in the Province of Nova Scotia, Canada," in Nova Scotia, Royal Commission on Provincial Development and Rehabilitation, *Report* (2 vols., Halifax, 1944), II, pp. 13, 48, 55-61; Kilbourn, *History of the Steel Company of Canada*, pp. 113-40.

25. Jones, "Gigantic Automaton," p. 61.

26. Author's interview with Harold Crowder, Sault Ste. Marie, 12 July 1984; Fitch, *Steel Workers*, pp. 6, 58; Craig Heron, Shea Hoffmitz, Wayne Roberts, and Robert Storey, *All That Our Hands Have Done: A Pictorial History of the Hamilton Workers* (Oakville, 1981), p. 43; AUCAI. Charles Rumford Walker had the same reaction when he first began work in an American steel plant in 1919; see Walker, *Steel*, pp. 8-10, 16-17.

27. OIFR, 1912, 1916; Nova Scotia, Factories Inspector, *Reports* (Halifax), 1912, Appendix 1 ("Report of Special Enquiry into One Hundred Accidents at Dominion Iron and Steel Co., Ltd., Sydney"), p. 3; 1916, p. 32; 1917, p. 6; Ontario Workmen's Compensation Board, *Reports* (Toronto, 1918-22); *Sault Star*, 16 August 1922.

28. Quoted in Ontario, Royal Commission on the Mineral Resources of Ontario and Measures for Their Development, *Report* (Toronto, 1890), p. 320. See also Sulte, *Forges*, p. 101; Stephens, "Boomtown," p. 25; Cameron, *New Glasgow*, pp. V1-V2, V8-V10; PANS, MG 3, Vol. 242, 21 March 1877, p. 232; *CMM*, 1879, p. 79; Andreae, "Iron Works," pp. 10-11, 14, 21, 23, 85.

29. *LG*, 1,5 (January, 1901), p. 223; 5,5 (November, 1904), p. 474; 6,11 (June, 1906), p. 1300; 7,2 (August, 1906), p. 112; 7,3 (September, 1906), pp. 226, 258; 7,5 (November, 1906), p. 459; 10,4 (October, 1909), p. 444; 11,2 (August, 1910), p. 175; 12,3 (September, 1911), p. 239; 13,3 (September, 1912), p. 235; 13,4 (October, 1912), p. 311; 14,1 (July, 1913), p. 12; 14,2 (August, 1913), p. 118; 14,5 (November, 1913), p. 523; 16,6 (December, 1915), p. 679; 16,11 (May, 1916), p. 1174; 16,15 (September, 1916), p. 1531; 16,16 (October, 1916), p. 1619; 16,17 (November, 1916), p. 1713; Sydney *Post*, 14 September 1906; *Evening News* (New Glasgow), 4, 7, 27 August 1915; 2, 19 August, 16 December 1916.

30. F.H. Bell, "Lifting and Conveying Material in the Foundry," *CF*, 12,3 (March, 1921), p. 19.

31. H.J. Habbakuk made this argument about American industry in his *American and British Technology in the Nineteenth Century: The Search for Labour-Saving Inventions* (Cambridge, 1962). His argument has created a large debate about the importance of labour scarcity in inducing technological change; see Peter Temin, "Labour Scarcity and the Problem of American Industrial Efficiency in the 1850s," *Journal of Economic History*, 26,3 (September, 1966), pp. 277-98; Temin, "Labor Scarcity in America," *Journal of Interdisciplinary History* (Winter, 1971); C.K. Harley, "Skilled Labour and the Choice of Technique in Edwardian Industry," *Explorations in Economic History*, 11,4 (Summer, 1974), pp. 391-414; Nathan Rosenberg, "The Direction of Technological Change: Inducement Mechanisms and Focusing Devices," *Economic Development and Cultural Change*, 18,1 (October, 1969), pp. 1-24; Rosenberg, *Technology and American Economic Growth* (New York, 1972); Paul David, *Technological Choice, Innovation, and Economic Growth: Essays on Amer-*

ican and British Experience in the Nineteenth Century (London, 1975), pp. 19-21.

32. Cameron, *New Glasgow*.

33. See, for example, AISI, *Yearbook* (New York, 1914), pp. 259-61.

34. *CMR*, 21,12 (December, 1902), p. 301.

35. NSCHLR, p. 66; Jones, "Gigantic Automaton," p. 118.

36. Ontario, Royal Commission on Mineral Resources, *Report*, pp. 323, 325, 331-35. Percentage calculations are mine. See also Steel Company of Canada, *Reports* (1874), p. 12, for an estimate of 13 per cent at Londonderry in 1874.

37. PAC, RG 87 (Mineral Resources Branch Records), Vol. 18, File 82. Calculations are mine.

38. See Charles More, *Skill and the English Working Class, 1870-1914* (London, 1980); Stephen Wood, ed., *The Degradation of Work?: Skill, Deskilling and the Labour Process* (London, 1982); Craig R. Littler, *The Development of the Labour Process in Capitalist Societies* (London, 1982); Ken C. Kusterer, *Know-How on the Job: The Important Working Knowledge of "Unskilled" Workers* (Boulder, Col., 1978); Ian Radforth, "Logging Pulpwood in Northern Ontario," in Heron and Storey, eds., *On the Job: Confronting the Labour Process in Canada*, pp. 248-53; Ian McKay, "Industry, Work, and Community in the Cumberland Coalfields, 1848-1927" (Ph.D. thesis, Dalhousie University, 1983), pp. 613-33; Ann Phillips and Barbara Taylor, "Sex and Skill: Notes Towards a Feminist Economics," *Feminist Review*, 6 (1980), pp. 79-88.

39. See, in particular, Andrew Zimbalist, ed., *Cases Studies in the Labor Process* (New York, 1979).

40. See *Capital* (Harmondsworth), 1 (1976), pp. 492-639; see also Nathan Rosenberg, "Marx as a Student of Technology," in Les Levidow and Bob Young, eds., *Science, Technology, and the Labour Process: Marxist Studies, Volume I* (London, 1981), pp. 8-31; and David Gartman, "Marx and the Labor Process: An Interpretation," *Insurgent Sociologist*, 8,2-3 (Fall, 1978), pp. 97-108.

41. Harry Braverman, *Labor and Monopoly Capitalism: The Degradation of Work in the Twentieth Century* (New York, 1974), p. 231.

42. Katherine Stone, "The Origins of Job Structures in the Steel Industry," *Radical America*, 7,6 (November-December, 1973), pp. 19-64. Michael Hanagan has applied the same perspective to the French steel industry; see *The Logic of Solidarity: Artisans and Industrial Workers in Three French Towns, 1871-1914* (Urbana, 1980), p. 132.

43. Bryan D. Palmer, *A Culture in Conflict: Skilled Workers and Industrial Capitalism in Hamilton, Ontario, 1860-1914* (Montreal, 1979); Craig Heron and Bryan D. Palmer, "Through the Prism of the Strike: Industrial Conflict in Southern Ontario, 1901-1914," *CHR*, 58,4 (December, 1977), pp. 423-58; Ian McKay, "Strikes in the Maritimes, 1901-1914," *Acadiensis*, 13,1 (Autumn, 1983), pp. 3-46; Craig Heron, "The Crisis of the Craftsman: Hamilton's Metal Workers in the Early Twentieth Century," *L/LT*, 6 (Autumn, 1980), pp. 7-48; David Montgomery, *Workers' Control in America: Studies in the History of Work, Technology, and Labor Struggles* (New York, 1979); Richard Edwards, "Social Relations of Production at the Point of Production," *Insurgent Sociologist*, 8,2-3 (Fall, 1978), pp. 109-25; David Stark, "Class Struggle and the Trans-

formation of the Labor Process: A Relational Approach," *Theory and Society*, 9,1 (January, 1980), pp. 89-130.

44. Wood, *Degradation of Work?*; More, *Skill*; Littler, *Labour Process*; William Lazonick, "Industrial Relations and Technical Change: The Case of the Self-Acting Mule," *Cambridge Journal of Economics*, 3 (1979); Wayne Lewchuk, "Fordism and British Motor Car Employers, 1896-1932," in Howard F. Gospel and Craig R. Littler, eds., *Managerial Strategies and Industrial Relations: An Historical and Comparative Study* (London, 1983), pp. 82-110; see also William H. Lazonick, "Technological Change and the Control of Work: The Development of Capital-Labour Relations in U.S. Mass Production Industries," *ibid.*, pp. 111-36.

45. Tony Manwaring and Stephen Wood, "The Ghost in the Machine: Tacit Skills in the Labor Process," *Socialist Review*, 74 (March-April, 1984), pp. 55-94.

46. Charles F. Sabel, *Work and Politics: The Division of Labor in Industry* (New York, 1982), pp. 59-70.

47. "How Mike Oleschuk Got His Farm," *CBM*, 27 (December, 1980), pp. 4-6; Walker, *Steel*; Williams, *What's On the Worker's Mind*, pp. 14-36; IWMR, p. 61; MacEachern interview; University of British Columbia Library, Special Collections, James Robertson Papers, Box 5, File 1, Handwritten Memorandum, p. 3; PANS, MG 2, Vol. 689 (E.H. Armstrong Papers), F10/18552, Dominion Iron and Steel Company Limited. In his 1910 book, John Fitch estimated the unskilled to be 60 per cent of the steelmaking work force in the Pittsburgh area; *Steel Workers*, p. 154.

48. P.W. Musgrove, *Technical Change, the Labour Force, and Education: A Study of the British and German Iron and Steel Industries, 1860-1964* (Oxford, 1967), pp. 25-68.

49. PANS, MG 1, Vol. 2155 (J. Dix Fraser, "Graham Fraser – His Life and Work"), pp. 91-92; Duncan Burn, *The Economic History of Steelmaking, 1870-1939* (Cambridge, 1961), pp. 10-11; OBMR, 1900, p. 101; William Smaill, "Iron Manufacture," *CE*, 4,1 (May, 1896), p. 14; *ibid.*, 10,11 (January, 1903), p. 19; BI, MG 14, 38 (Dominion Iron and Steel Papers), 2 (Chief Chemist's Letterbooks), Charles A. Arnold to David Baker, 12 August 1903; Williams, *Peter White*, p. 179; Bernard McEvoy, *From the Great Lakes to the Wide West: Impressions of a Tour Between Toronto and the Pacific* (Toronto, 1902), pp. 30-31; "Sault Ste. Marie and its Industries," *CE*, 10,1 (January, 1903), p. 19; Halifax *Herald*, 22 October 1922; *CM*, 21,26 (26 June 1919), p. 656. On the general phenomenon of turning science into a direct tool of capitalist production, see David Noble, *America By Design: Science, Technology, and the Rise of Corporate Capitalism* (New York, 1977).

50. Fitch, *Steel Workers*, pp. 39-44; "Jim Hines: A View From the Open Hearth," *CBM*, 27 (December, 1980), pp. 13-17; J.D. Fraser, "From Iron Ore to Steel: A Sketch of Iron Mining and Manufacturing in Pictou County, Nova Scotia," *CE*, 2,5 (September, 1894), p. 135; Jones, "Gigantic Automaton," p. 102; Walker, *Steel*, pp. 34-37; Williams, *What's On the Worker's Mind*, pp. 35-36.

51. Cameron, *New Glasgow*, p. V10; RCRCL, *Evidence – Nova Scotia*, p. 265.

52. Fitch, *Steel Workers*, pp. 48-53; Wilson and Theurkauf, "What the Eyes Behold," p. 339; "Wally Chandler: Catching Steel," *CBM*, 27 (December, 1980),

p. 19; Wayne Roberts, ed., *Baptism of a Union: Stelco Strike of 1946* (Hamilton, 1981), p. 16. See also Harry Jack Waisglass, "A Case Study in Union-Management Co-operation" (M.A. thesis, University of Toronto, 1948), p. 116; Williams, *What's On the Worker's Mind*, pp. 41-70.

53. George MacEachern, "Autobiography" (typescript), p. 31; "Lew Allen Davis & the Railroad," *CBM*, 28 (June, 1981), pp. 9-11; Crowder interview.

54. "Eight-Hour Committee," p. 166.

55. NSCHLR, p. 68; PANS, MG 2, Vol. 689, F10/18552, Dominion Iron and Steel Company Limited; MacEachern, "Autobiography," p. 31.

56. "Jim Hines," p. 15.

57. Jones, "Gigantic Automaton," p. 102.

58. Especially Braverman, *Labor and Monopoly Capital*; Stone, "Job Structures."

59. United States, Bureau of Labor, *Report of Employment in the Iron and Steel Industry* (4 vols., Washington, 1911-13), III, p. 81. For more recent discussions, see More, *Skill*; Sabel, *Work and Politics*, pp. 57-70.

60. AISI, *Yearbook*, 1919, p. 414.

61. BI, MG 14, 38, 3, M. Shiras to W.W. McKeown, Jr., 22 April 1901; to L.W. Squire, 17 June 1901; and to C. McCrery, 18 June 1901; J.H. Means to David Baker, 10 August 1901 and 2 October 1901.

62. *Sault Star*, 19 June 1902, 7 March 1907; *CE*, 13,2 (February, 1906), p. 42.

63. After examining British evidence, Charles More concludes similarly that, "although casually acquired the skill required [in semi-skilled work] was genuine and in some cases considerable, and might arise from quite a lengthy acquaintance with the work." *Skill*, p. 130.

64. John H. Ashworth, *The Helper and American Trade Unions* (Baltimore, 1915); IWMR, p. 132.

65. *Conciliation Court*, pp. 17-18, 26-27, 32-36, 81.

66. *Ibid.*, pp. 18-19, 33, 36, 83.

67. See, for example, "Lew Allan Davis & the Railroad," p. 9; *SF*, 14,6 (June, 1950); AUCAI. On the general phenomenon of internal labour markets, see Peter B. Doeringer and Michael J. Piore, *Internal Labor Markets and Manpower Analysis* (Lexington, Mass., 1971); Paul Osterman, ed., *Internal Labor Markets* (Cambridge, Mass., 1984).

68. One striking example of this process occurred in 1920 when Disco borrowed the head roller from a steel plant in Homestead, Pennsylvania, to train local men to run the corporation's new plate mill. According to *Iron and Steel of Canada* (May, 1920, p. 111), "no one watching the operations of rolling, shearing and stocking at the present time would suspect that nearly all of the workmen were quite new to the work."

69. Patrick McGeown described how job ladders in a British steel plant locked men into one department in one plant, just as they undoubtedly did in Canada. (McGeown, *Heat the Furnace*, p. 8.) John Fitch discovered the same patterns in Pittsburgh. (*Steel Workers*, p. 13.)

70. PANS, MG 1, 169 (Thomas Cantley Letterbooks), 1903-07, 506 (Thomas Cantley to A.L. Lang, 1 February 1905); author's interview with Earl Orchard, Sault Ste. Marie, 12 July 1984; "C.M. (Clem) Anson and Steel," *CBM*, pp. 43-57; Crowder interview; Stone, "Job Structures," p. 51.

71. Cameron, *New Glasgow*; L. Anders Sandberg, "The Closure of the Ferrona Iron Works, 1904," *Acadiensis*, 14,1 (Autumn, 1984), pp. 98-104; *CE*, 11,11 (November, 1904), p. 347; RCIRE, III, p. 2299; *Sault Star*, 22 September 1904; see also AUCAI; MacEachern, "Autobiography"; *Sault Star*, 27 April 1921, 6 December 1922; *Financial Post*, 24 August 1928; Orchard interview. As Charles Sabel has argued, semi-skilled workers move with their company or return to old jobs primarily because of their fear of falling back into the unskilled labour market outside the factory; see *Work and Politics*, pp. 96-98.

Chapter 3

1. Donald, *Canadian Iron and Steel Industry*, p. 209.
2. Cameron, *New Glasgow*, pp. V1, V8-V9; BI, MG 14, 38, 3 (Dominion Iron and Steel Company, Blast Furnace Superintendent's Letterbooks), J.H. Means to D. Baker, 30 September 1901.
3. Sydney *Post*, 4, 5 July 1901, 17 November 1903; Sydney *Record*, 16 November 1903; *Sault Star*, 16, 30 January, 6 March 1902, 21 July, 13 October 1904, 6 April 1905, 27 July 1911; Labour Canada Library (Hull), Provincial Workmen's Association, Grand Council, Minutes, III, September, 1904, pp. 417-18; *LG*, 4,5 (November, 1904), p. 474; Frances M. Heath, "Labour, the Community, and Pre-World War I Immigration Issue," in *50 Years of Labour in Algoma*, pp. 44-56; Ronald F. Crawley, "Class Conflict and the Establishment of the Sydney Steel Industry, 1899-1904" (M.A. thesis, Dalhousie University, 1980).
4. *Sault Star*, 23 January 1902; Saint John *Sun*, 10 August 1901.
5. Cameron, *New Glasgow*; Fritz, *Autobiography*, pp. 60, 128; Bernard Elbaum, "The Making and Shaping of Job and Pay Structures in the Iron and Steel Industry," in Paul Osterman, ed., *Internal Labor Markets* (Cambridge, Mass., 1984), pp. 78-82. At the end of World War I, Algoma would go so far as to offer special night classes, which would enable the corporation "to promote its own men to higher positions when vacancies occur." *Sault Star*, 7 February, 14 June, 30 October, 8 November 1919; *LG*, 21,1 (January, 1921), p. 7.
6. BI, MG 14, 38, 3, S.M. Shiras to F.W. Waterman, 20 December 1900; Shiras to A. Graham, 28 December 1900; Sydney *Post*, 13 July 1904; *Conciliation Court*, pp. 11, 18, 26-27, 29, 31-36, 50, 60-63; A.R. Kennedy, "Position of Canada's Steel and Iron Industry," *IC*, January, 1919, pp. 163-65.
7. MacEachern, "Autobiography"; AUCAI. In 1931 census takers found only sixty-nine machinists' apprentices, alongside 1,267 machinists, and only fourteen moulders' apprentices, alongside 412 moulders in the Canadian steel industry. *Census of Canada*, 1931, VII, p. 814.
8. Crawley, "Class Conflict," pp. 52-53; *Conciliation Court*, p. 63; *Evening News* (New Glasgow), 5 May, 6 September 1917; RCIRE, New Glasgow, p. 3589; Joy Parr, "Hired Men: Ontario Agricultural Wage Labour in Historical Perspective," *L/LT*, 15 (Spring, 1985), pp. 99-100; John MacDougall, *Rural Life in Canada: Its Trends and Tasks* (Toronto, 1973 [1913]); Alan A. Brookes, "Out-Migration from the Maritime Provinces, 1860-1900: Some Preliminary Considerations," in P.A. Buckner and David Frank, eds., *Atlantic Canada After Confederation* (Fredericton, 1985), pp. 34-63; Patricia A. Thornton, "The Prob-

lem of Out-Migration from Atlantic Canada, 1871-1921: A New Look," *Acadiensis*, 15, 1 (Autumn, 1985), pp. 3-34; Kari Levitt, *Population Movements in the Atlantic Provinces* (Halifax, 1960).

9. *CMR*, 16,2 (February, 1897), p. 52; see also C.A. Meisner, "Notes on Some Comparisons Between Southern and Nova Scotia Iron Methods," *CMR*, 16,1 (January, 1897), pp. 12-15.

10. Emilia Kolcon-Lach, "Early Italian Settlement at Sault Ste. Marie, Ontario, 1898-1921" (M.A. thesis, University of Western Ontario, 1979), pp. 3-4, 19-20, 38; Aileen Collins Hinsperger, *Stories of the Past: 300 Years of Soo History* (n.p., n.d. [1967]), p. 60; J. Konarek, "Algoma Central and Hudson's Bay Railway: The Beginnings," *Ontario History*, 62,2 (June, 1970), p. 77; Heath, "Immigration Issue," pp. 41-42; Robert F. Harney, "Montreal's King of Italian Labour: A Case Study of Padronism," *L/LT*, 4 (1979), pp. 57-84.

11. PANS, MG 1, Vol. 1191 C (Thomas Cozzolino, "Autobiography"), pp. 29-32; BI, Reports: Ethnic, E.J. Julian, "Brief History of the Italian Colony of Cape Breton"; Reports: Sydney, William G. Snow, "An Historical Sketch of Whitney Pier," p. 5; Tape 372 (Tony Bruno); MG 14, 38, 3, 15 December 1900 – 28 November 1901; *LG*, 1,8 (April, 1901), p. 389; Sydney *Post*, 21 December 1901, 25 February 1902; Crawley, "Class Conflict," pp. 60-61.

12. BI, Reports: Sydney, Steve Melnick, "A Family History of the Melnyks-Melnicks," p. 2; Reports: Ethnic, Imelda Gillis, "Ukrainian and Black Communities," pp. 19-21; Snow, "Whitney Pier," p. 11; MGTC (Canon George A. Francis), "History of the Black Population at Whitney Pier"; John Huk, "The Ukrainians of Sydney," Cape Breton *Post*, 27 July 1960; Ralph Wayne Ripley, "Industrialization and the Attraction of Immigrants to Cape Breton County, 1893-1914" (M.A. thesis, Queen's University, 1980), pp. 53-65; *Evening News* (New Glasgow), 15 April 1914; Watson Kirkconnell, "Kapuskasing – An Historical Sketch," *Queen's Quarterly*, 28,3 (January-March, 1921), p. 269; Desmond Morton, *The Canadian General: Sir William Otter* (Toronto, 1974), p. 343; Donald Avery, *"Dangerous Foreigners": European Immigrant Workers and Labour Radicalism in Canada, 1896-1932* (Toronto, 1979), p. 32; *CLL*, 8 June 1918; RCIRE, Sydney, pp. 3918-19; Carmela Patrias, "Patriots and Proletarians: The Politicization of Hungarian Immigrants in Canada, 1924-1946" (Ph.D. thesis, University of Toronto, 1985), pp. 170-79.

13. *LG*, 1,5 (January, 1901), p. 223.

14. Kilbourn, *History of the Steel Company of Canada*, pp. 121, 124.

15. Cozzolino, "Autobiography"; Robert F. Harney, "The Commerce of Migration," *Canadian Ethnic Studies*, 9,1 (1977), pp. 42-53; Harney, "Montreal's King of Italian Labour"; BI, Tape 187 (Roman Srivak); Tape 372 (Tony Bruno); Reports: Ethnic, Stephanie Melnick, "The Sydney Polish Community," pp. 1-8; *LN*, 30 October 1914; Paul Body, "Emigration from Hungary, 1880-1956," in N.F. Dreisziger *et al.*, eds., *Struggle and Hope: The Hungarian-Canadian Experience* (Toronto, 1982), pp. 37-38; Anthony W. Rasporich, *For a Better Life: A History of the Croatians in Canada* (Toronto, 1982), pp. 29-51, 56; Heron, "Working-Class Hamilton," pp. 288-378.

16. *LG*, 4,4 (October, 1903), p. 291; Nova Scotia, Secretary of Industries and Immigration, *Reports* (Halifax); Snow, "Whitney Pier," p. 4; Crawley, "Class Con-

flict," p. 59; *Census of Canada*, 1921, II, p. 363; PANS, MG 1, no. 174 (Thomas Cantley Papers), Correspondence, 1886-1900, H.J. Townshend to Thomas Cantley, 22 June 1900; Sydney *Post*, 13, 22, 31 May, 15, 17 June 1901; *Evening News* (New Glasgow), 15 April 1914; Peter Neary, "Canadian Immigration Policy and the Newfoundlanders, 1912-1939," *Acadiensis*, 11,2 (Spring, 1982), pp. 69-83; Neary, *Bell Island: A Newfoundland Mining Community, 1895-1966* (Canada's Visual History, Series 1, No. 12, Ottawa, 1974).

17. NSCHLR, p. 69; Hamilton *Spectator*, 3 March 1919; *Census of Canada*, 1911, II, pp. 208-09, 372, 374; 1921, I, pp. 542, 544; 1931, II, pp. 495, 498-99. This particular ethnic mix departs somewhat from the pattern in American steel plants, where Newfoundlanders were unknown, Italians played a more limited role, and Slovaks were much more prominent. In general the American steel work force also had a higher percentage of European-born and second-generation Europeans. United States, Immigration Commission, *Reports: Immigrants in Industries* (42 vols., Washington, 1911), *Vol. 8, Part 2: Iron and Steel Manufacturing (vol. I)*, p. 33.

18. NSCHLR, p. 69; PAC, RG 27 (Department of Labour), Vol. 294, File: "Reports – Employment Offices (B-N)," Hamilton; MacEachern, "Autobiography," p. 41; MacEachern interview; Storey, "Workers, Unions, and Steel," p. 129; Matthew James Foster, "Ethnic Settlement in the Barton Street Region of Hamilton, 1921 to 1961" (M.A. thesis, McMaster University, 1965), pp. 140, 142; *LG*, 18,3 (March, 1918), p. 180.

19. Harney, "Commerce of Migration"; Harney, "Montreal's King of Italian Labour"; Robert F. Harney, "Men Without Women: Italian Migrants in Canada, 1885-1930," *Canadian Ethnic Studies*, 11,1 (1979), pp. 29-47; Harney, "The Padrone System and Sojourners in the Canadian North, 1885-1920," in George E. Pozzetta, ed., *Pane e Lavore: The Italian American Working Class* (Toronto, 1980), pp. 119-37; Body, "Emigration from Hungary"; Henry Radecki and Benedyckt Heydenkorn, *A Member of a Distinguished Family: The Polish Group in Canada* (Toronto, 1976), pp. 26-28; John-Paul Himka, "The Background to Emigration: Ukrainians of Galicia and Bukovyna, 1848-1914," in Manoly R. Lupul, ed., *A Heritage in Transition: Essays in the History of Ukrainians in Canada* (Toronto, 1982), pp. 11-31; Wsevolod W. Isajiw, "Occupational and Economic Development," *ibid.*, pp. 59-84; Rasporich, *For a Better Life*, pp. 29-74; Carmela Patrias, "Hungarian Immigration to Canada Before the Second World War," *Polyphony*, 2,2-3 (1979-80), pp. 17-44; Avery, *"Dangerous Foreigners"*, pp. 16-38; Robert F. Foerster, *The Italian Immigration of Our Times* (New York, 1919), pp. 64-105; John Bodnar, *Immigration and Industrialization: Ethnicity in an American Mill Town, 1870-1940* (Pittsburgh, 1977); Frank Thistlethwaite, "Migration From Europe Overseas in the Nineteenth and Twentieth Centuries," in Herbert Moller, ed., *Population Movements in Modern European History* (New York, 1964), pp. 73-92; Robert Eugene Johnson, *Peasant and Proletarian: The Working Class of Moscow in the Late Nineteenth Century* (New Brunswick, New Jersey, 1979).

20. *Sault Star*, 10 April 1902, 18 April 1907, 19 March 1912.

21. Sydney *Post*, 16, 21 May, 15, 16 June 1901, 14 September 1906; Neary, "Canadian Immigration Policy," pp. 69-83.

22. Bryce M. Stewart, "The Housing of Our Immigrant Workers," Canadian Political Science Association, *Papers and Proceedings*, 1913, p. 98; Jane Synge, "Immigrant Commmunities – British and Continental European – in Early Twentieth Century Hamilton, Canada," *Oral History*, 4,2 (Autumn, 1976), p. 42; Kolcon-Lach, "Italian Settlement," pp. 34-38; *Sault Star*, 30 November 1911, 14 July 1914; Hamilton *Spectator*, 27 October 1906; Hamilton *Herald*, 12 November 1913.

23. See Stephen Castles and Godula Kosack, *Immigrant Workers and the Class Structure in Western Europe* (London, 1973); John Berger and Jean Mohr, *A Seventh Man* (Harmondsworth, 1975); Michael J. Piore, *Birds of Passage: Migrant Labor and Industrial Societies* (London, 1979).

24. *Census of Canada*, 1911, II, pp. 426, 430, 434; 1921, II, pp. 348, 361, 363; 1931, II, pp. 746, 754, 756.

25. Ripley, "Industrialization," pp. 36-40; Melnick, "Sydney Polish Community," p. 11; undated *Sault Star* clipping in Sault Ste. Marie Public Library, Reference Department, Scrapbook 34; Van Emery, "Clergue," p. 202.

26. Hamilton *Herald*, 5 June 1911, 31 August 1912; Hamilton *Spectator*, 17 June 1904, 18 October 1905, 27 October 1906; "The Housing Situation in Hamilton," *Canadian Municipal Journal*, 8,7 (July, 1912), pp. 255-56; Melnick, "Family History," pp. 4-5; Robert F. Harney, "Boarding and Belonging," *Urban History Review*, 2-78 (October, 1978), pp. 8-37; Patrias, "Patriots and Proletarians," pp. 179-84.

27. Methodist Church, Department of Temperance and Moral Reform, and Presbyterian Church, Board of Social Service and Evangelism, *Report of a Preliminary and General Social Survey of Hamilton* (n.p., n.d. [1913]); *Sydney, Nova Scotia: The Report of a Brief Investigation of Social Conditions in the City . . .* (n.p., n.d. [1913]); Stewart, "Housing," pp. 106-09; see also BI, Tape 187 (Roman Srivak); *Sault Star*, 18 March 1921; MacEachern interview; Sydney *Post*, 16, 31 May, 6, 17 June 1901. In 1911 the men made up 57.3 per cent of Sydney's Newfoundland population. *Census of Canada*, 1911, II, p. 430.

28. See Edmund W. Bradwin, *The Bunkhouse Man: A Study of Work and Pay in the Camps of Canada, 1903-1914* (Toronto, 1972 [1928]).

29. Sydney *Post*, 24 November 1905 (see also 6 June 1901); Hamilton *Spectator*, 18 October 1905; Kolcon-Lach, "Italian Settlement," pp. 102-03.

30. *Sault Star*, 10 April 1902. For a more hostile journalistic comment on the same subject, see *ibid.*, 5 March 1903.

31. Hamilton *Spectator*, 15 February 1919. For a statement of similar sentiments from Disco, see Avery, *"Dangerous Foreigners"*, p. 67.

32. Hamilton *Spectator*, 8 April 1907, 15 February 1919; Hamilton *Herald*, 18 February 1919; Sydney *Post*, 28 October 1918; "Eight-Hour Committee," p. 178; PAC, RG 27, Vol. 294, File: "Reports – Employment Offices (B-N)," Hamilton; RG 36, 11, Vol. 1, File 2-3, "Iron and Steel Wages; Comparisons – 1928, America and Europe"; Canada, Board of Inquiry into Cost of Living, *Report* (2 vols., Ottawa, 1915), II, p. 142; *Census of Canada*, 1921, II, p. 456; IV, p. 401; *LG*, 16,5 (November, 1915), p. 620; 21,3 (March, 1921), p. 472; 30,12 (December, 1930), p. 1468; Canada, Department of Labour, *Wages and Hours of Labour in Canada, 1920-1929* (Ottawa, 1930), p. 38; BI, MG 14, 38, 3, J.H.

Means to B. Bryan, 28 August 1901; RCIRE, New Glasgow, pp. 3567, 3638, 3836-37; CBMR, p. 539.

33. *ISC*, 3,1 (February, 1920), p. 34; RCIRE, p. 3832; PAC, RG 36/11 (Advisory Board on Tariff and Trade), Vol. 1, File 2-1, Exhibit 10; PANS, MG 2, Vol. 4 (E.H. Armstrong Papers), F13/1229.

34. Avery, *"Dangerous Foreigners"*, pp. 90-141; John Herd Thompson, "Bringing in the Sheaves: The Harvest Excursionists, 1890-1929," *CHR*, 59,4 (December, 1978), pp. 467-89.

35. Hamilton *Herald*, 15 August, 29 September 1916; 8, 11 January, 24, 25 April, 13, 30 July 1917; *CLL*, 10 August 1918; RCIRE, Sydney, pp. 3768, 3832; BI, Tape 8841 (Gloria McDougall), Tape 799 (Ed Parris), Melnick, "Sydney Polish Community," p. 9.

36. PAC, MG 30, A 16 (Sir Joseph Flavelle Papers), Vol. 3, File 26, pp. 17-25; RG 36/11, Vol. 8, File 3-6, Exhibit 16 ("Report to Dominion of Canada Advisory Tariff Commission Board as Concerning Employees of Algoma Steel Corporation Limited, Sault Ste. Marie, Ontario"); *Census of Canada*, 1931, VII, pp. 904-05, 912, 919, 940-41, 958-59, 966-67, 974-75; 1941, VII, pp. 896-97, 904-05. Calculation of naturalized citizens is mine.

37. Hamilton *Times*, 10, 11, 14 April 1902; Hamilton *Spectator*, 11 April 1902, 18 October 1905, 8 April 1907; Kilbourn, *History of the Steel Company of Canada*, pp. 121, 124; PAC, RG 27 (Department of Labour), Vol. 299, File 3475.

38. NSCHLR, pp. 64-75.

39. William I. Thomas and Florian Znaniecki, *The Polish Peasant in Europe and America* (5 vols., New York, 1918-20), I, p. 199; RCIRE, p. 3651.

40. RCIRE, p. 3744. See also *LG*, 1,8 (April, 1901); Sydney *Post*, 6 April 1901.

41. PAC, RG 27, Vol. 299, File 3475. See Provincial Workmen's Association, Grand Council, Minutes, III, September, 1904, pp. 442, 448.

42. Avery, *"Dangerous Foreigners"*, pp. 65-89; *CLL*, 25 February 1919 (see also 29 December 1917, 12 January, 30 March, 18 May, 1 June 1919); Hamilton *Spectator*, 11 February 1919; see also *EE*, 22 March, 17 May, 12 July 1919; *Amalgamated Journal*, 2 August 1917; Kolcon-Lach, "Italian Settlement," pp. 128-30; Julian Sher, *White Hoods: Canada's Ku Klux Klan* (Vancouver, 1983), pp. 27-30; *Worker* (Toronto), 12 October 1929; Clifford Rose, *Four Years with the Demon Rum, 1925-1929: The Autobiography and Diary of Temperance Inspector Clifford Rose*, ed., E.R. Forbes and A.A. MacKenzie (Fredericton, 1980), p. 77; Rob Bostelaar, "Sault Was Divided at Gore Street," *Sault Star*, 19 March 1976; Bostelaar, "Klan Tarred and Feathered One Man," *ibid.*, 20 March 1976; Morley Torgov, *A Good Place to Come From* (Toronto, 1974), p. 16; AUCAI; *Sault Star*, 13 April 1920.

43. See W. Peter Ward, *White Canada Forever: Popular Attitudes and Public Policy Toward Orientals in British Columbia* (Montreal, 1978); William Tuttle, *Race Riot: Chicago in the Red Summer of 1919* (New York, 1970).

44. On Hamilton, see Foster, "Ethnic Settlements"; Diana Brandino, "The Italians of Hamilton, 1921-1945" (M.A. thesis, University of Western Ontario, 1977), pp. 62-63, 97, 102-03; Enrico Cumbo, "Italians in Hamilton, 1900-40," *Polyphony*, 7,2, pp. 28-36; William Boleslaus Makowski, *History and Integration of Poles in Canada* (Lindsay, Ont., 1967), pp. 75-76, 96; Father Camillo Lando,

"Italian National Catholic Churches of Hamilton," in Mary G. Campanella, ed., *Proceedings of Symposium 1977* ... (np. [Hamilton], 1977), pp. 67-68; Susan M. Papp, "Hungarians in Ontario," *Polyphony*, 2,2-3 (1979-80), p. 8; Hamilton *Herald*, 11 June 1910, 8 May, 13-14 June, 23 September 1911, 29 October 1912, 29 March, 12 June 1913, 14 June 1915, 23, 25, 30 April 1917; Hamilton *Times*, 29 October 1917. On Sault Ste. Marie, see *Sault Star*, 17 October 1901, 5 April 1913, 13 April 1920; John R. Cameletti, *A History of the Separate Schools of the City of Sault Ste. Marie* (Sault Ste. Marie, 1967), pp. 8-9. On Sydney, see Sydney *Post*, 25 February 1902, 15, 17 August 1908, 4 September 1918; BI, Gillis, "Ukrainian and Black Communities," pp. 3-11, 22-24; Sister Mary Anne Morrison, "The Ukrainians of Cape Breton," pp. 4-5; MG 7F, 1 (Anna Kiec, "History of St. Mary's Polish Parish"); Tape 187 (Roman Srivak), Tape 351 (Winston Ruck), Tape 414 (Helen Blazey), Tape 747 ("The Road to Whitney Pier"); Esperanza Maria Razzolonin, *All Our Fathers: The North Italian Colony in Industrial Cape Breton* (Halifax, 1983).

45. See, for example, *Sault Star*, 28 November 1901, 9, 16 January, 10 April, 23 October 1902, 10 January 1907, 11 February 1914; Sydney *Post*, 8-9 August, 3, 14, 18 June 1901, 10 October 1902, 4 May 1903, 16, 19 July 1904, 10 December 1907, 28 October, 2 November 1911; Halifax *Herald*, 3 August 1920; Ripley, "Industrialization," pp. 41-42.

46. See, for example, *CLL*, 19 January 1918; Torgov, *Good Place*, p. 22.

47. Hamilton *Herald*, 4 April 1910.

48. David M. Gordon, Richard Edwards, and Michael Reich, *Segmented Work, Divided Workers: The Historical Transformation of Work in the United States* (New York, 1982).

49. Probably the most pessimistic view of these ethnic divisions is presented in Gabriel Kolko's *Main Currents in Modern American History* (New York, 1976), pp. 67-99.

50. *Sault Star*, 5 September 1901; Hamilton *Herald*, 18 November 1912; Hamilton *Times*, 7 May 1912.

51. Nova Scotia, Factories Inspector, *Report* (Halifax), 1912, p. 8; 1917, p. 8.

52. In *Steelworkers in America: The Nonunion Era* (New York, 1960), David Brody argues that this divided work force, with its ethnically and occupationally segmented structure, was a "source of stability" for the corporations. In Canada at least (and perhaps in the United States as well), that stability was often elusive, as the steelmaking firms struggled to keep a regular supply of appropriate unskilled labour available and to keep the internal job ladders working adequately. As we will see in Chapter Four, the craftsmen and labourers regularly proved to be thorns in the sides of steel-plant managers.

53. Richard Edwards employs the term (mistakenly, in my view) to the new factory regime of the early twentieth century in the United States. See *Contested Terrain: The Transformation of the Workplace in the Twentieth Century* (New York, 1979).

54. Fitch, *Steel Workers*, pp. 166-70; IWMR, pp. 44-84; "Eight-Hour Committee," pp. 168-69, 171, 174. Many steelworkers' reminiscences convey the same sense that the twelve-hour day was all-consuming. See, for example, *SF*, 14,6 (June, 1950); AUCAI; MacEachern, "Autobiography"; Walker, *Steel*, pp. 62-80, 143-

57. Whiting Williams, who worked these long shifts in American steel plants in 1919, described the mental state that resulted as a "chronic listlessness which is not exactly fatigue but comes from the two-days-off-in-two-years habit"; see *What's On the Worker's Mind*, p. 31.

55. AUCAI; *Sault Star*, 18 April 1907; RCIRE, Sydney, pp. 3739, 3754-55, 3780; *AJ*, 19 February 1920, p. 27; Sydney *Post*, 18 January 1929; MacEachern, "Autobiography," pp. 95-97; *Conciliation Court*, pp. 13-14; Cameron, *New Glasgow*, pp. V5-V6; McCracken, "Steel Industry," pp. 268-70; *LG*, 30,1 (January, 1930), p. 2; Social Service Council of Canada, Committee on Industrial Life, *A Survey of the Steel Industry in Canada* (Toronto, 1930); Charles Hill, "Fighting the Twelve-Hour Day in the American Steel Industry," *Labor History*, 15,1 (Winter, 1974), pp. 19-35; Brody, *Steelworkers*, pp. 270-75; International Labour Organization, *The Application of the Three-Shift System to the Iron and Steel Industry* (Geneva, 1922).

56. Elbaum and Wilkinson, "Industrial Relations and Uneven Development" pp. 283-93; Elbaum, "Job and Pay Structures," pp. 71-108; Montgomery, *Workers' Control in America*, pp. 11-12; David A. McCabe, *The Standard Rate in American Trade Unions* (Baltimore, 1912), pp. 62-65; Steven R. Cohen, "Steelworkers Rethink the Homestead Strike of 1892," *Pennsylvania History*, 48,2 (April, 1981), pp. 159-63; McGeown, *Heat the Furnace*, pp. 32, 59.

57. RCRCL, *Evidence – New Brunswick*, pp. 17-18, 107; *Evidence – Nova Scotia*, pp. 237-68, 388-407; *Evidence – Ontario*, pp. 760-64, 786-94; Cameron, *New Glasgow*, pp. V1-V6; *SF*, 14,6 (June, 1950), p. 20.

58. RCRCL, *Evidence – Nova Scotia*, p. 247

59. "Eight-Hour Committee," p. 188. See also RCRCL, *Evidence – Nova Scotia* pp. 237, 240, 242, 249, 257-59, 266, 268, 390, 394-95, 398; *Evidence – Ontario* pp. 760, 763, 786-87.

60. Brody, *Steelworkers*, pp. 27-29; *Conciliation Court*, pp. 60-61; University of British Columbia Library, Special Collections, James Robertson Papers, Box 5, File 1, "Notes from Conversations with Officers of the Steel Company of Canada, Hamilton, Ont., Dec.21/23," p. 1. The American data indicate that this new wage policy allowed the steel companies to cut the earnings of skilled steelworkers dramatically (see Brody, *Steelworkers*, pp. 27-49); comparable data for Canada are not available.

61. PANS, MG 3, Vol. 242 (Steel Company of Canada, Council of Administration, Minutes, 1876-77), 24 January 1877, p. 182; 14 March 1877, p. 218; 21 March 1877, p. 233; Steel Company of Canada, Ltd., *The Twenty-Fifth Milestone, 1910-1935: A Brief History of Stelco* (Hamilton, 1935), p. 12.

62. Brody, *Steelworkers*, pp. 27-49; BI, MG 14, 38, 3; Kilbourn, *History of the Steel Company of Canada*, pp. 84-85; *Sault Star*, 10 August 1905; Crawley, "Class Conflict," p. 74; Craig Heron, "Punching the Clock," *Canadian Dimension*, 14,3 (December, 1979), pp. 26-29. Scotia's instruction manual for keeping "cards of account" can be found in DUA, MG 4, p. 106.

63. Stone, "Job Structures," pp. 19-64.

64. Daniel Nelson, *Frederick W. Taylor and the Rise of Scientific Management* (Madison, Wisc., 1980).

65. BI, MG 14, 38, 3, G. Shiras to H.L. Burnham, 19 January 1901; *LG*, 9,5

(November, 1908), p. 468; Sydney *Post*, 3, 5, 6, 9, 10 October 1908; NSCHLR, p. 70; *Sault Star*, 11 November 1908; "Eight-Hour Committee," pp. 177, 183, 185-86, 189; *Conciliation Court*; RCIRE, New Glasgow, pp. 3566-67; *Amplified Brief of the Algoma Steel Corporation, Limited* (n.p., n.d.), p. 14; Robertson Papers, "Conversations"; IWMR, p. 129; Robert Hessen, "The Transformation of Bethlehem Steel, 1904-1909," *Business History Review*, 46,3 (Autumn, 1972), pp. 355-56. John Fitch's investigations into piecework in American steel plants led him to conclude that "when the rate is judiciously cut from time to time, the tonnage system of payment becomes the most effective scheme for inducing speed that has yet been devised." *Steel Workers*, p. 189. Like Fitch, Michael Burawoy has emphasized, on the basis of his study of a post-World War II machine shop, how piecework can focus workers' frustrations on each other, since any delays can cut into larger earnings; see Burawoy, *Manufacturing Consent: Changes in the Labor Process Under Monopoly Capitalism* (Chicago, 1979).

66. *LG*, 5,11 (May, 1905), p. 1074; 6,7 (January, 1906), p. 703; 6,11 (May, 1906), p. 1225; 6,12 (June, 1906), p. 1336; 7,3 (September, 1906), p. 258; 7,4 (October, 1906), p. 367; 7,6 (December, 1906), p. 945; 7,9 (March, 1907), p. 939; 8,1 (July, 1907), p. 13; 9,4 (October, 1908), p. 359; 10,6 (December, 1909), p. 634; 9,5 (November, 1910), p. 534; *CM*, 10,7 (14 August 1913), p. 170; Ferris, *Algoma's Industrial and Trade Union Development*, pp. 40-45, 61; *ISC*, 13,3 (March, 1930); *Sault Star*, 19 January, 16 February, 30 March, 20 April, 12 October 1905, 15 February, 28 June, 12 July, 9 August, 13 September 1906, 14 November 1907, 18 November 1909.

67. AUCAI. The same phenomenon was described at the Sydney steel plant by George MacEachern in his "Autobiography," p. 39, and in American plants by John Fitch, *Steel Workers*, pp. 185-88; and IWMR, p. 129.

68. "Eight-Hour Committee"; *Conciliation Court*; RCIRE.

69. RCIRE, New Glasgow, pp. 3534-35; MacEachern, "Autobiography," p. 28; AUCAI.

70. Montgomery, *Workers' Control*, pp. 9-31; Gregory S. Kealey, *Toronto Workers Respond to Industrial Capitalism, 1867-1892* (Toronto, 1980), pp. 37-97.

71. On the general phenomenon of the "drive system," see Fitch, *Steel Workers*, p. 185; Sumner H. Slichter, "The Management of Labor," *Journal of Political Economy*, 27,10 (December, 1919), pp. 815-16n; and Slichter, *The Turnover of Factory Labor* (New York, 1919), pp. 202-03, 375; Daniel Nelson, *Managers and Workers: Origins of the New Factory System in the United States, 1880-1920* (Madison, 1975), pp. 43-44.

72. See, for example, BI, MG 14, 38, 3.

73. Tamara K. Hareven found the same arrangements between the employment office and overseers in the Amoskeag textile mills in New Hampshire; see Hareven, *Family Time and Industrial Time: The Relationship Between the Family and Work in a New England Industrial Community* (Cambridge, 1982), pp. 42-43. It seems likely that Daniel Nelson has exaggerated the erosion of foremen's power in his influential study, *Managers and Workers*. See Sanford M. Jacoby, *Employing Bureaucracy: Managers, Unions, and the Transformation of Work in American Industry* (New York, 1985).

74. Sydney *Post*, 17-18 November 1903; MacEachern interview, and "Autobiography," pp. 27, 42; AUCAI; Roberts, *Baptism*, p. 11; Storey, "Workers, Unions, and Steel," p. 212.

75. Hamilton *Herald*, 4 April 1910; *Sault Star*, 30 November 1921, 30 May 1922; *AJ*, 20 January 1921, p. 18; MacEachern interview; AUCAI; Roberts, *Baptism*, p. 11; "How Mike Oleschuk Got His Farm," *CBM*, 27 (December, 1980), p. 5; Waisglass, "Case Study," p. 93; Fitch, *Steel Workers*, pp. 143-44.

76. Sydney *Post*, 18 November 1903, 25, 30 July, 1 August 1918; *Conciliation Court*, p. 61; RCIRE, Sydney, p. 3741; "The 1923 Strike in Steel and the Miners' Sympathy Strike," *CBM*, 22 (June, 1979), p. 3.

77. BI, Tape 52; Waisglass, "Case Study," p. 92.

78. RCIRE, III, p. 2316; *LG*, 21,3 (March, 1921), p. 522; *Sault Star*, 8, 28, 29 November 1919, 7 February 1920.

79. PAC, RG 27, Vol. 69, File 222(7), W.L. King to W. Mulock, 22 July 1904 (in which King indicates Disco has a list of twenty-eight it would not re-employ after the 1904 strike); BI, Tape 52; Tape 532 (Carl Neville); AUCAI; MacEachern interview, and "Autobiography," pp. 58-59; *MLH*, 6 May, 11 November 1922, 17 February, 10 November 1923.

80. PAC, MG 30, A 16, Vol. 2, File 11 (Department of Labour), R. Hobson to J.W. Flavelle, 8 July 1916; PANS, MG 2, Vol. 666, File 8; 670, File 5; BI, Tape 158 (George MacEachern); MacEachern, "Autobiography," pp. 37, 120-21; PAO, RG 23 (Ontario Provincial Police Records), Series E (Criminal Investigations), 30, 1.6 (Strikes and Agitations, 1913-1921).

81. RCIRE, Sydney, pp. 3793-96.

82. *MLH*, 11 November 1922.

83. See, in particular, Stone, "Job Structures," pp. 40-43. On the general phenomenon of job ladders, see Peter B. Doeringer and Michael J. Piore, *Internal Labor Markets and Manpower Analysis* (Lexington, Mass., 1971).

84. Elbaum, "Job and Pay Structures," pp. 74-82. See also Fitch, *Steel Workers*, pp. 141-42; Williams, *What's on the Worker's Mind*, p. 21; IWMR, p. 129; Sabel, *Work and Politics*, pp. 59-63.

85. See United States, Bureau of Labor, *Report on Conditons of Employment in the Iron and Steel Industry* (4 vols., Washington, 1911-13); United States, Commissioner of Corporations, *Report on the Steel Industry* (3 vols., Washington, 1911-13); United States, Immigration Commission, *Reports: Immigrants in Industries, Iron and Steel Manufacturing*; Fitch, *Steel Workers*.

86. RCIRE, Hamilton, p. 2289.

87. Steel Company of Canada, *Reports Upon the Property at Londonderry*, p. 81; Andreae, "Iron Works," pp. 50, 190-93; Birkinbine, "Old and New Iron Industry Compared," p. 90.

88. Pentland, *Labour and Capital in Canada*, pp. 24-26.

89. Fraser, "Graham Fraser"; Cameron, *New Glasgow*; Sandberg, "Deindustrialization," pp. 192-95.

90. See, for example, the concerned letters of Disco's blast furnace superintendent in BI, MG 14, 38, 27 May, 5 September 1901.

91. On Disco, see *CMM*, 1901, p. 128; Crawley, "Class Conflict," pp. 54, 56; *CBMR*, p. 527; NSCHLR, p. 70; *LG*, 13,8 (February, 1913), p. 816; RCIRE,

Sydney, pp. 3831, 3834; *ISC*, 10,11 (November, 1927), p. 341. On Scotia, see PANS, MG 1, 1309 (Nova Scotia Steel and Coal Company Limited, 1909), p. 1. On Algoma, see *Sault Star*, 17 October 1901, 13 April 1920.

92. Sydney *Post*, 4 March 1903; Crawley, "Class Conflict," p. 124; RCIRE, Sydney, p. 3820.

93. Scotia's was founded in 1889, Algoma's in 1901, Stelco's in 1902, and Disco's in 1906. *Conciliation Court*, pp. 73, 76-77, 82-83; Halifax *Herald*, 18 January 1921; Fraser, "Graham Fraser," pp. 80-81; Sydney *Post*, 9 April, 22 November 1906, 13 December 1916; *LG*, 6,12 (June, 1906), p. 1301; 7,7 (January, 1907), pp. 783-84; 7,12 (June, 1907); 9,1 (July, 1908), pp. 70-71; 9,12 (June, 1909), p. 1344; 11,12 (June, 1911), p. 1344; 12,8 (February, 1912), p. 725; 12,12 (June, 1912), p. 1135; Nova Scotia, House of Assembly, *Journals and Proceedings* (Halifax), 1912, Pt. 2, App. 15 (Factory Inspector, *Report*), App. 1, p. 1; Ontario, Inspector of Insurance and Registrar of Friendly Societies, *Report* (Toronto), 1902, p. C-130; 1915, pp. 132-33, 148-49; *LG*, 21,3 (March, 1921), p. 37; AUCAI; Sandberg, "Deindustrialization," p. 194.

94. *ISC*, 2,11 (December, 1919), p. 311.

95. DUA, MG 4, 106, General Correspondence, 1919-25, File: B.D. McKnight; File: Knight & Knight; File: Town of New Glasgow, A. McCoul to A. McColl, 25 November 1923; AUCAI.

96. AISI, *Monthly Bulletin*, 2 (1914), p. 341; Sir William Ralph Meredith, *Final Report on Laws Relating to the Liability of Employers to Make Compensation to Their Employees for Injuries Received in the Course of Their Employment* . . . (Toronto, 1913), pp. 32-33, 36; PAC, MG 31, B3 (C.H. Speer, "Algoma Steel Corporation"), p. 122; RG 36/8, Vol. 6, File 13 (Sydney), 1955; *History of the Medical Profession* (Sault Ste. Marie, 1922), pp. 33, 43, 51; *Evening News* (New Glasgow), 31 July 1918.

97. Hamilton *Herald*, 28 April 1911; *LG*, 14,2 (August, 1913), p. 117; RCIRE, Hamilton, pp. 2289, 2299; *ISC*, 6,3 (April, 1923), p. 68; Steel Company of Canada, *Annual Reports* (Hamilton), 1925-26.

98. See Ross Harkness, *J.E. Atkinson of the Star* (Toronto, 1963), pp. 134-53; R.M MacIver, *Labor in the Changing World* (Toronto, 1919); J.O. Miller, ed., *The New Era in Canada* (London, 1917); William Irvine, *The Farmers in Politics* (Toronto, 1976 [1920]); Louis Aubrey Wood, *A History of Farmers' Movements in Canada* (Toronto, 1975 [1924]); Gregory S. Kealey, "1919: The Canadian Labour Revolt," *L/LT*, 13 (Spring, 1984), pp. 11-44; Craig Heron, "Labourism and the Canadian Working Class," *ibid.*, pp. 45-77; Salem Bland, *The New Christianity* (Toronto, 1973 [1920]); Richard Allen, *The Social Passion: Religion and Social Reform in Canada, 1914-28* (Toronto, 1973), pp. 63-80.

99. Stelco, *Annual Report*, 1919, n.p.; *ISC*, July, 1924, p. 129; March, 1928, p. 68; *LG*, 26,3 (March, 1926), pp. 237-39; 28,2 (February, 1928), pp. 172-73; 28,9 (September, 1928), pp. 942-43.

100. *LG*, 28,9 (September, 1928), p. 943.

101. *Sault Star*, 4 September, 22 October 1920; J.L. Charlesworth, "The Company Paper's Part in Improving Industrial Relations," *IC*, 21,8 (December, 1920), p. 84; *BB*, 1,1 (February, 1925), p. 1; Canada, Department of Labour, *Employees' Magazines in Canada* (Ottawa, 1921), pp. 3, 13, 21.

102. On Algoma's sports program, see almost every issue of the *Sault Star* after the war; on the Steel Plant Club, see *ibid.*, 16, 31 January, 9, 19, 24 March, 9, 7, 13, 15 April, 26 July 1920, 18 April 1921; author's interview with Earl Orchard; AUCAI; on Algoma's "industrial classes," see *Sault Star*, 7 February, 14 June, 30 October, 8 November 1919; *LG*, 21,1 (January, 1921), p. 7.

103. *ISC*, 3,7 (August, 1920), p. 208.

104. Brody, *Steelworkers*, pp. 164-68.

105. *LG*, 22,3 (March, 1922), p. 263; 22,10 (October, 1922), pp. 1041-42; 30,7 (August, 1930), p. 783; *ISC*, 4,7 (August, 1921), pp. 198-200; 5,7 (August, 1922), p. 147; *Sault Star*, 20 June, 17, 24 November 1921, 4 January, 25 February, 16 August 1922; on Stelco, see Kilbourn, *History of the Steel Company of Canada*, p. 123; *LG*, 23,11 (November, 1923), p. 1271; on Disco, see *ISC*, August, 1920, p. 208; September, 1922, pp. 168-69; January, 1928, p. 24; March, 1928, p. 88; March, 1929, p. 72; *BB*, 56 (27 February 1926), p. 4; *LG*, 26,11 (November, 1926), p. 1106; 29,1 (January, 1929), pp. 45-46; 29,4 (April, 1929), pp. 398-99; 29,5 (May, 1929), p. 505; 30,5 (May, 1930), p. 544.

106. *IC*, 21,4 (August, 1920), p. 60; *Sault Star*, 20 June 1921, 25 February 1922.

107. *Sault Star*, 24 November 1921, 16 August 1922.

108. *LG*, 18,9 (September, 1918), p. 696; 21,3 (March, 1921), pp. 521-22; *IC*, 21,11 (March, 1921). Both Massey-Harris and International Harvester used their industrial councils for the same purposes in the early 1920s; see Bruce Scott, "'A Place in the Sun': The Industrial Council at Massey-Harris, 1919-1929," *L/LT*, 1 (1976), pp. 172-76; and Heron, "Working-Class Hamilton, 1895-1930," p. 100.

109. *CLL*, 4 May 1918; *EE*, 9 August 1919; *LG*, 23,1 (January, 1923), p. 5; 26,7 (July, 1926), pp. 665-66; *BB*, 10 (11 April 1925), pp. 1-2; Canada, Commission to Inquire into the Industrial Unrest Among the Steel Workers at Sydney, N.S., *Report* (Ottawa, 1924), pp. 16-19.

110. RCIRE, Hamilton, p. 2290; Hamilton *Herald*, 17 May 1920; Storey, "Workers, Unions, and Steel," pp. 199-213.

111. *BB*, 10 (11 April 1925), p. 2.

112. PANS, MG 2, Vol. 593 (E.N. Rhodes Papers), pp. 14525-26; MG 3, 1877, p. 49; Halifax *Herald*, 23 September 1921; *BB*, 38 (24 October 1925), p. 1; Sydney *Post*, 4 April 1930; PAC, RG 36/8 (Tariff Enquiry and Commission), Vol. 6, File 13, 1955-63; RG 36/11, Vol. 1, File 2-1, Exhibit 10; File 3-6, Exhibit 16; McDowall, *Steel at the Sault*, p. 175.

113. Jacoby, *Employing Bureaucracy*. The other major study in disagreement with Brody is Stuart D. Brandes, *American Welfare Capitalism, 1880-1940* (Chicago, 1976).

114. Canada, Dominion Bureau of Statistics, *Iron and Steel and Their Products* (Ottawa, 1924), p. 45.

115. Peter Warrian interview with Ted Barbet; Storey, "Workers, Unions, and Steel"; Storey, "Unionization versus Corporate Welfare: The 'Dofasco Way,'" *L/LT*, 12 (Autumn, 1983), pp. 7-42.

116. Gerald Zahavi, "Negotiated Loyalty: Welfare Capitalism and the Shoeworkers of Endicott Johnson, 1920-1940," *Journal of American History*, 70,3

(December, 1983), pp. 602-20; MacEachern, "Autobiography," pp. 57-58, 91-97; BI, William G. Snow, "Sydney Steelworkers: Their Troubled Past and the Birth of Lodge 1064" (typescript), pp. 18-19; MG 19, Doane Curtis, "Akes and Pains of the Labor Strugal," pp. 11, 14-18; PAC, RG 36/11, Vol. 1, File 2-2; H.A. Logan, "Report on Labour Relations," in Nova Scotia, Royal Commission on Provincial Development and Rehabilitation, *Report*, II, p. 82n; Social Service Council of Canada, *Survey of the Steel Industry*; McCracken, "Steel Industry," pp. 269-70; Sydney *Post*, 17 July 1930; Storey, "Workers, Unions, and Steel," pp. 199-213, 308-38.

117. PAO, RG 7 (Ministry of Labour), VII-1, Vol. 8 (Research Branch, Senior Investigator, 1926-29); Orchard interview; Storey, "Workers, Unions, and Steel."

Chapter 4

1. A.W. Macdonald, "Labor Turnover of Industrial Plants, and What Steps Can Be Taken to Minimize It," *ISC*, 3,4 (May, 1920), p. 123; G.W. Austen, "Excessive Labor Turnover and its Remedies," *IC*, 21,5 (May, 1920), p. 74.

2. Slichter, *Turnover of Factory Labor*; Paul Frederick Brissenden and Emil Frankel, *Labor Turnover in Industry* (New York, 1922), pp. 36-37; Ronald W. Schatz, *The Electrical Workers: A History of Labor at General Electric and Westinghouse* (Urbana, 1983), p. 17; Stephen Meyer III, *The Five Dollar Day: Labor Management and Social Control in the Ford Motor Company, 1908-1921* (Albany, 1981), p. 80; United States, Bureau of Labor, *Report on Conditions of Employment in the Iron and Steel Industry*, III. In the first meetings of American personnel managers before the war, there was a consensus "that the workmen rarely stayed long in their plants and that it was necessary constantly to hire men in large numbers in order to maintain the working force on an even level." Paul H. Douglas, "Plant Administration of Labor," *Journal of Political Economy*, 27,7 (July, 1919), pp. 544-60.

3. See, for example, Hamilton *Times*, 6 April 1907, 7 May 1912; Hamilton *Herald*, 18 November 1912, 12 November 1913; Sydney *Post*, 18, 31 May, 9 August, 21 December 1901, 24 November 1905; *Sault Star*, 24 August, 12 September 1901, 24 April, 24 July 1902, 5 March, 23 July, 6 August 1903, 15 September 1904, 18 April 1907; *Evening News* (New Glasgow), 15 August 1913, 15 April 1914; *LG*, 1,8 (April, 1901), p. 389; 4,7 (January, 1904), p. 615.

4. NSCHLR, p. 69; Kolcon-Lach, "Early Italian Settlement," pp. 34-38; *Conciliation Court*, p. 62; Sandberg, "Deindustrialization," p. 162.

5. The problem of increased turnover during the war affected virtually all industries across the continent. "The tremendous increase in the demand for labor caused by the war has resulted in a pronounced increase in the size of turnover and has impressed manufacturers with the intimate relation between the stability of labor and the state of the labor market," Slichter wrote in *Turnover of Factory Labor*, p. 29. See also R.H., "The Industrial Slacker," *CF*, 9,5 (May, 1918), p. 105; PAC, MG 26, H 1(c) (Sir Robert Borden Papers), Vol. 211, p. 18813 (C.B. Gordon to General Elliot, 18 August 1916); MG 30, A 16 (Sir Joseph

Flavelle Papers), Vol. 2, File 11, T. Findley to T. Crothers, 31 March 1916; Vol. 38, File 1918-19, M. Irish to J. Flavelle, 20 June 1918.

6. Nova Scotia, Factories' Inspector, *Report* (Halifax), 1917, p. 6; PAC, MG 26, H 1(c), Vol. 211, p. 118826 (D.H. McDougall to Mark Workman, 2 September 1916); Sydney *Post*, 14 September 1918; RCIRE, Sydney, p. 3831.

7. Hamilton *Spectator*, 8 August 1916; see also Hamilton *Herald*, 12 August 1916.

8. PAC, MG 26, H 1(c), Vol. 216, p. 121857 (W.C. Franz to J.W. Flavelle, 20 January 1917); *LG*, 18,3 (March, 1918), p. 180.

9. Sydney *Post*, 19 July 1918; Hamilton *Herald*, 6, 10, 13, 19 April 1918; *Sault Star*, 11, 15 May 1918; *LN*, 2 August 1918. The same pattern of absenteeism and reduced effort appeared in World War II; see Logan, "Report on Labour Relations," II, pp. 9-10.

10. Transiency was not a new theme in North American working-class life. Several historians have established the nineteenth-century patterns of great geographical mobility; see, for example, Michael B. Katz, *The People of Hamilton, Canada West: Family and Class in a Mid-Nineteenth-Century City* (Cambridge, Mass., 1975); Stephan Thernstrom and Peter R. Knights, "Men in Motion: Some Data and Speculations about Urban Population Mobility in Nineteenth-Century America," in Tamara K. Hareven, ed., *Anonymous Americans: Explorations in Nineteenth Century Social History* (Englewood Cliffs, N.J., 1971); Paul G. Faler, *Mechanics and Manufacturers in the Early Industrial Revolution: Lynn, Massachusetts, 1780-1860* (Albany, 1981), pp. 140-43.

11. *LG*, 7,12 (June, 1907), p. 1371; 8,3 (September, 1907), p. 279; 8,7 (January, 1908), p. 914; 10,4 (October, 1909), p. 444; 10,7 (January, 1910), p. 769; 10,8 (February, 1910), p. 874; 10,12 (June, 1910), p. 1351; 11,1 (July, 1910), p. 9; 12,7 (January, 1912), p. 633; 13,2 (August, 1912), p. 124; 14,6 (December, 1913), p. 651; 14,9 (March, 1914), p. 1020; 14,12 (June, 1914), p. 1375; *Sault Star*, 22 August 1907; *MLH*, 9 June 1923.

12. PAC, RG 27 (Department of Labour), Vol. 304, File 16 (27A).

13. *LG*, 3,7 (January, 1903), p. 487; 4,3 (September, 1903), p. 185; 8,8 (February, 1908), p. 887; 8,9 (March, 1908), p. 1039; 8,11 (May, 1908), p. 1311; 9,1 (July, 1908), p. 13; 9,3 (September, 1908), p. 253; 14,7 (January, 1914), p. 757; 14,10 (April, 1914), p. 1127; 14,11 (May, 1914), p. 1253; 15,3 (September, 1914), pp. 339-40; 15,4 (October, 1914), p. 449; 15,5 (November, 1914), p. 559; 15,6 (December, 1914), p. 664; 15,7 (January, 1915), p. 764; 15,10 (April, 1915), p. 1143; 15,11 (May, 1915), p. 1260; *ISC*, 2,7 (August, 1919), p. 193; 3,3 (April, 1921), p. 67; *EE*, 22 March 1919; *MLH*, 13 May 1922; PAC, RG 36/11, Vol. 1, File 2-1, Exhibit 10; Vol. 8, File 3-6, Exhibit 16; PANS, MG 2, Vol. 4 (E.A. Armstrong Papers), F13/1229.

14. *Conciliation Court*, p. 62; see also Hamilton *Times*, 6 April 1907.

15. On itinerant labourers, see Edmund Bradwin, *The Bunkhouse Man* (Toronto, 1972); A. Ross McCormack, *Reformers, Rebels, and Revolutionaries: The Western Canadian Radical Movement, 1899-1919* (Toronto, 1977), pp. 98-117; John Herd Thompson, "Bringing in the Sheaves: The Harvest Excursionists, 1890-1929," *CHR*, 59,4 (December, 1978), pp. 467-89; Jack London, *The Road* (London, 1967); G.H. Westbury, *Misadventures of a Working Hobo in Canada*

(Toronto, 1930); Carleton H. Parker, *The Casual Laborer and Other Essays* (New York, 1920), pp. 61-89.

16. Macdonald, "Labor Turnover," p. 123; Austen, "Excessive Labor Turnover," p. 74; see also Brissenden and Frankel, *Labor Turnover*, pp. 96-98.

17. William Davenport, "As a Britisher Sees It," *Western Clarion*, 30 June 1906.

18. Kolcon-Lach found many references to dismissals for strike activity in Algoma's pre-1910 payroll book, but the *Labour Gazette's* Sault Ste. Marie correspondent reported none of them. On the 1903 riot, see Ducin, "Labour's Emergent Years and the 1903 Riots," pp. 1-14.

19. Crawley, "Class Conflict," pp. 70-74; Heron, "Working Class Hamilton," pp. 343-48.

20. Crawley, "Class Conflict," pp. 72-75; Sydney *Post*, 2-4 March 1903; Ian McKay, "Strikes in the Maritimes, 1901-1914," *Acadiensis*, 13,1 (Autumn, 1983), p. 23.

21. The following account is based on Craig Heron, "Hamilton Steelworkers and the Rise of Mass Production," Canadian Historical Association, *Historical Papers*, 1982, pp. 122-24.

22. See, for example, Eric Hobsbawm, *Primitive Rebels: Studies in Archaic Forms of Social Movements in the 19th and 20th Centuries* (New York, 1959); Hobsbawm, "Peasants and Politics," *Journal of Peasant Studies*, 1,1 (1973), pp. 3-22; Robert Eugene Johnson, *Peasant and Proletarian: The Working Class of Moscow in the Late Nineteenth Century* (New Brunswick, N.J., 1979), pp. 159-62.

23. Robert F. Harney, "Ethnicity and Neighbourhoods," in Harney, ed., *Gathering Place: Peoples and Neighbourhoods of Toronto, 1834-1945* (Toronto, 1985), pp. 16-17; Patrias, "Patriots and Proletarians," pp. 187-93, 273-75, 292-96; Hamilton *Herald*, 18 November 1912; Avery, *"Dangerous Foreigners"*, pp. 39-64; Orest T. Martynowych, "Village Radicals and Peasant Immigrants: The Social Roots of Factionalism among Ukrainian Immigrants in Canada, 1896-1918" (M.A. thesis, University of Winnipeg, 1978); Peter Krawchuck, *The Ukrainian Socialist Movement in Canada (1907-1918)* (Toronto, 1979); Sydney *Post*, 3 March 1903; Hamilton *Spectator*, 1 April 1910.

24. Hamilton *Spectator*, 1 April 1910; Hamilton *Times*, 2 April 1910.

25. See Wayne Roberts, "Artisans, Aristocrats, and Handymen: Politics and Unionism among Toronto Skilled Building Trades Workers, 1896-1914," *L/LT*, 1 (1976), pp. 92-121; "Toronto Metal Workers and the Second Industrial Revolution, 1889-1914," *ibid.*, 6 (Autumn, 1980), pp. 49-72; Roberts, "The Last Artisans: Toronto Printers, 1896-1914," in Gregory S. Kealey and Peter Warrian, eds., *Essays in Canadian Working Class History* (Toronto, 1976), pp. 125-42; Heron and Palmer, "Through the Prism of the Strike," pp. 423-58; Heron, "Crisis of the Craftsman," pp. 7-48; Montgomery, *Workers' Control in America*, pp. 91-112.

26. Crawley, "Class Conflict," pp. 64-71; see also Heron, "The Crisis of the Craftsman," pp. 35, 37-40; Myer Siemiatycki, "Munitions and Labour Militancy: The 1916 Hamilton Machinists' Strike," *L/LT*, 3 (1978), pp. 131-52; BI, 12, 139, c, 19 March 1913.

27. PAC, RG 27, Vol. 300, File 3620; *Machinists' Monthly Journal*, 25 (1913), pp. 45-46, 254-55; *Sault Star*, 11, 27 November 1912. John Fitch reported in 1910 that American steelmaking corporations would not under any circumstan-

ces stop Sunday work: "Workmen told me that it was easier to get excused on any other night than Sunday. In some of the mills men are discharged if they refuse to work on Sunday." *Steel Workers*, p. 176.

28. Hamilton *Spectator*, 2, 4-5 April 1900; Hamilton *Times*, 2-3 April 1900; *CE*, 7 (1899-1900), p. 332; Crawley, "Class Conflict," p. 64; Sandberg, "Deindustrialization," p. 159; Sydney *Post*, 3, 5 October 1908; *Eastern Chronicle* (New Glasgow), 17 March 1908; PAC, RG 27, Vol. 304, File 15 (23); Vol. 306, File 17(47); MG 26, H 1 (c) (Sir Robert Borden Papers), Vol. 216, pp. 121849-61; *Sault Star*, 17, 19, 20-22 July 1915, 29-30 September 1916, 16 January 1917.

29. Robert H. Babcock, *Gompers in Canada: A Study in American Continentalism Before the First World War* (Toronto, 1974), pp. 38-54.

30. Sandberg, "Closure of the Ferrona Iron Works," p. 99

31. Crawley, "Class Conflict," pp. 70-78, 86-89; Joe McDonald, "Unionism of the Steel Plant," Miners' Museum (Glace Bay, N.S.), *Museum Notes*, 2 (July, 1977), n.p.; Sydney *Post*, 11, 17, 18, 21, 28, 30 November, 2, 4 December 1903; Sydney *Record*, 1 October, 11, 16, 18, 20, 23, 30 November, 1, 2 December 1903 (clippings in BI, MG 12, 24, f, 1).

32. Crawley, "Class Conflict," pp. 90-131; McDonald, "Unionism"; PAC, RG 27 (Department of Labour), Vol. 69, File 222(7); Sydney *Post*, 1-4, 11 June, 6, 13, 18, 23 July 1903.

33. *Conciliation Court*, p. 8.

34. *Ibid.*, pp. 2-10; *Evening News* (New Glasgow), 27 August, 14 November, 24 December 1913; 24 June, 20, 23 July 1915. For the tensions around rate-setting in car plants, see DUA, MG 4, 106 (Hawker-Siddeley Papers), General Correspondence, 1919-25, File: Dominion Coal General, A.McColl to D.H. McDougall, 18 April 1923; PAC, RG 27, Vol. 303, File 14(8); Vol. 323, File 20; Vol. 343, File 29(87).

35. *Evening News* (New Glasgow), 7 November 1913, 6 July 1914; *Machinists' Monthly Journal*, 26,11 (December, 1914), p. 1170; Rose, *Four Years with the Demon Rum*, pp. 5-9, 83.

36. See *Conciliation Court* for all the documents and testimony presented to the hearing, which review all the developments of 1915; also *Evening News* (New Glasgow), 28 January, 3 April, 11, 16, 22 June, 13, 14, 16-18, 20, 23, 27 August, 21, 30 September 1915; PAC, RG 27, Vol. 304, File 15 (28); *LG*, 15,3 (September 1915), p. 287; 15,4 (October, 1915), pp. 441-45.

37. *Evening News* (New Glasgow), 2, 7, 26 September, 2, 5-6, 8, 11, 13, 17, 23 November 1915, 5 January, 21 March, 5, 25 April, 1, 26-27 May, 20 July, 6 September, 18 November, 1, 9 December 1916, 20 February, 12, 15, 21, 28, 30 March, 5 April 1917; PAC, RG 27, Vol. 304, File 15 (42); Vol. 306, File 17 (52), File 17 (60). Four years later a writer in the Halifax *Herald* (17 February 1919) reached the cynical conclusion, "When the employees had got their wages up as high as they could hope for, their interest in the union began to wane and the union began to fall to pieces." The same story noted the personal rivalry that had grown up between Dane and some leaders of the local Independent Labour Party.

38. BI, American Federation of Labor, *Membership Book and Working Card*, held by James A. Casher from December, 1916; *Evening News* (New Glasgow), 21

April 1917; Sydney *Post*, 2 November 1917, 6 February 1918; *AJ*, 19,15 (27 December 1917), p. 4; *CLL*, 26 February 1918.

39. Jesse E. Robinson, *The Amalgamated Association of Iron, Steel, and Tin Workers* (Baltimore, 1920); Brody, *Steelworkers*, pp. 1-79; Arthur G. Burgoyne, *The Homestead Strike of 1892* (Pittsburgh, 1979 [1893]); Palmer, *A Culture in Conflict*, pp. 83-85; *LG*, 2,4 (October, 1901), pp. 248-49.

40. *CLL*, 9-30 March, 6, 13 April, 4 May, 1-29 June, 6-27 July, 3-24 August 1918; *AJ*, 18, 25 April 1918; *Evening News* (New Glasgow), 6, 14 March, 1, 3, 15, 17-19 April, 1 May, 5, 12, 13-15, 17, 18, 20, 24 June 1918; *Eastern Chronicle* (New Glasgow), 16, 19 April 1918; Sydney *Post*, 2, 4-8 March, 27, 30 April, 25-28 June, 11, 19, 20, 22-25, 27, 29, 30 July, 1-2, 10, 12, 15, 17, 19, 20, 22 August 1918; Halifax *Herald*, 17 June, 29, 30 July 1918; PAC, MG 30, A 16 (Sir Joseph Flavelle Papers), Vol. 14, File 146 (Nova Scotia Steel and Coal Co. Strike 1918); RG 27, Vol. 308, File 18(77), File 18(85); David Frank, "The Cape Breton Coal Miners, 1917-1926" (Ph.D. thesis, Dalhousie University, 1979); RCIRE, Sydney, p. 3747.

41. PAC, MG 26, H 1 (c), Vol. 216, p. 121857 (W.C. Franz to J.W. Flavelle, 20 January 1917); *AJ*, 4, 25 January, 15 February, 8 March, 10, 24, 31 May, 5, 12, 26 July, 2, 9 August, 8, 24 November 1917, 9 May 1918; *LG*, 17,9 (September, 1917), p. 686; 17,10 (October, 1917), p. 790; 18,3 (March, 1918), pp. 177-81; 18,4 (September, 1918), p. 709; *Sault Star*, 6 November 1918; AUCI.

42. Craig Heron and George de Zwaan, "Industrial Unionism in Eastern Ontario: Gananoque, 1918-21," *Ontario History*, 77,3 (September, 1985), pp. 159-82; *AJ*, 3 April, 24 July, 11 September, 16 October 1919; *LN*, 23 August, 13 September 1918, 5 September, 3 October 1919; Hamilton *Spectator*, 24 September 1919.

43. *AJ*, 12 April 1917; *CLL*, 13 April 1918.

44. A union leader told the Royal Commission on Industrial Relations that the Sydney lodge's 2,564 members (in a work force of some 4,000) were "practically all trained men" (exclusive of the craft unionists). RCIRE, Sydney, p. 3738; M.T. Montgomery, "Stelco Story," United Steelworkers of America, *Information*, August-September, 1954, p. 5; McMaster University Archives, M.T. Montgomery Papers, Interview; AUCAI; *AJ*, 8 January, 5, 12 February 1920. The Hamilton lodge also had members in the local Burlington Steel plant, and there were lodges at independent rolling mills in Redcliffe, Alta., Selkirk, Man., London, Ont., Toronto, and Dartmouth, N.S. *LN*, 12 March 1920; *AJ*, 8 March 1917, 5 September 1918, 5 June 1919, 8 July, 16 December 1920.

45. RCIRE, Sydney, p. 3744; see also pp. 3753, 3756-57.

46. Montgomery interview; *AJ*, 11, 25 March, 20 May, 26 August 1920; *LN*, 30 September 1920; Hamilton *Times*, 25 October 1920; David Brody, *Labor in Crisis: The Steel Strike of 1919* (Philadelphia, 1965). Separate meetings were held in Whitney Pier as early as March, 1919, but a separate lodge, dominated by "foreigners," was not created until the fall of 1922. *AJ*, 6 March 1919, 9 November 1922. It is also possible that, in contrast to the United States, where Europeans had been integrated into the steel work force much earlier, and where a second generation was already emerging in the mills with aspirations to promotion in the steel plants (IWMR, pp. 129,135), more of the Europeans in

Canadian steel plants were still tied to sojourning and therefore less interested in the new unions.

47. Siemiatycki, "Munitions and Labour Militancy"; Heron, "Working-Class Hamilton"; Canada, Department of Labour, *Labour Organization in Canada* (Ottawa, 1911-19).

48. Siemiatycki, "Munitions and Labour Militancy"; D.J. Bercuson, "Organized Labour and the Imperial Munitions Board," *Relations industrielles/Industrial Relations*, 28 (1973), pp. 602-16; Jeffrey Gordon Hucul, "Canada and the International Labour Organization in the Interwar Period, 1919-1940" (M.A. thesis, University of Windsor, 1983), pp. 61-67; Brody, *Steelworkers*, pp. 199-213; Robert D. Cuff, "The Politics of Labor Administration During World War I," *Labor History*, 21,4 (Fall, 1980), pp. 546-69.

49. Paul Craven, *"An Impartial Umpire": Industrial Relations and the Canadian State, 1900-1911* (Toronto, 1980); Ben M. Selekman, *Postponing Strikes: A Study of the Industrial Disputes Investigation Act of Canada* (New York, 1927), pp. 168-81; *Industrial Banner*, 7 April 1916 (quoted in Siemiatycki, "Munitions and Labour Militancy," p. 137); *LG*, 17,12 (December, 1917), p. 984.

50. See Gregory S. Kealey, "1919: The Canadian Labour Revolt," *L/LT*, 13 (Spring, 1984), pp. 11-44; Craig Heron, "Labourism and the Canadian Working Class," *ibid.*, pp. 66-70; Heron and De Zwaan, "Industrial Unionism in Eastern Ontario"; Bryan D. Palmer, *Working-Class Experience: The Rise and Reconstitution of Canadian Labour, 1800-1980* (Toronto, 1983), pp. 170-77; see also James E. Cronin, "Labor Insurgency and Class Formation: Comparative Perspectives on the Crisis of 1917-1920 in Europe," in James E. Cronin and Carmen Sirianni, eds., *Work, Community, and Power: The Experience of Labor in Europe and America, 1900-1925* (Philadelphia, 1983), pp. 2-17; and David Montgomery, "New Tendencies in Union Struggles and Strategies in Europe and the United States, 1916-1922," *ibid.*, pp. 88-116.

51. The major concerns to be discussed here parallelled those being raised by American steelworkers at this point; see the list of demands produced by the National Committee to Organize Steel Workers, reproduced in William Z. Foster, *The Great Steel Strike and Its Lessons* (New York, 1920), p. 77.

52. See, for example, *CLL*, 3 August 1918, 16 March 1918; RCIRE, New Glasgow, p. 3611, Sydney, pp. 3741-42, 3749.

53. *CLL*, 25 January 1919; *EF*, 22 March, 17 May 1919.

54. See *CLL*, 13 April, 3, 24 August 1918.

55. *Evening News* (New Glasgow), 4 May 1917, 14 June 1918; *CLL*, 13 July 1918; *WW*, 14 November 1919; Sydney *Post*, 30 July, 1 August 1918; RCIRE, Sydney, p. 3741.

56. *Evening News* (New Glasgow), 29 June 1916; *WW*, 4 June 1920; see also RCIRE, New Glasgow, pp. 3587-88, 3665-66; *WW*, 14 November 1919.

57. Sydney *Post*, 19, 23 July, 26 October 1918; *CLL*, 3, 24 August 1918; *WW*, 2 July 1918; *AJ*, 31 October 1918; *LG*, 20,7 (July, 1920), pp. 831-33.

58. *AJ*, 15 February, 12 July 1917.

59. *Ibid.*, 13 April 1918.

60. *AJ*, 22 July 1920, p. 16.

61. *CLL*, 9 March, 10 August, 12 October 1918.

62. RCIRE, New Glasgow, pp. 3533-55.

63. *AJ*, 17 July 1919.

64. *CLL*, 4 March 1920.

65. *Ibid.*, 18 March 1920; *WW*, 3 September 1920.

66. These are discussed in Montgomery, *Workers' Control in America*, pp. 9-31; Gregory S. Kealey, " 'The Honest Workingman' and Workers' Control: The Experience of Toronto's Skilled Workers, 1860-1892," *L/LT*, 1 (1976), pp. 32-68.

67. Sydney *Post*, 17 November 1903; Sumner H. Slichter, *Union Policies and Industrial Management* (Washington, 1941), p. 1.

68. See, for example, *CLL*, 3 August 1918, where the new president of Scotia was assessed as knowing "how to treat our class of people"; and *MLH*, 6, 13 May 1922, where it is argued that as a result of Besco's managerial incompetence "the technical processes of production were neglected."

69. RCIRE, New Glasgow, p. 3535.

70. *EF*, 29 March 1919; RCIRE, New Glasgow, pp. 3534-36; Sandberg, "Deindustrialization," pp. 182-92.

71. Canada, National Industrial Conference of Dominion and Provincial Governments with Representative Employers and Labour Men, on the Subjects of Industrial Relations and Labour Laws, and for the Consideration of the Labour Features of the Treaty of Peace, *Official Report of Proceedings and Discussions Together with Various Memoranda Relating to the Conference and the Report of the Royal Commission on Industrial Relations* (Ottawa, 1919), p. 89.

72. See H.L. Harris, *The Right to Manage*.

73. *CLL*, 4 May, 24 August 1918; Sydney *Post*, 11 July 1918.

74. For a discussion of the ideology of these parties, see Heron, "Labourism and the Canadian Working Class," pp. 45-76.

75. David Frank, "Company Town/Labour Town: Local Government in the Cape Breton Coal Towns, 1917-1926," *Histoire sociale/Social History*, 14,27 (May, 1981), pp. 177-96; MacEwan, *Miners and Steelworkers*, pp. 65-71; *WW*, 30 July 1920; C.D. Martin, "Algoma Labour Becomes Politically Active, 1914-1922," in *50 Years of Labour in Algoma*, pp. 57-76; Heron, "Working-Class Hamilton," pp. 619-48; Heron and De Zwaan, "Industrial Unionism in Eastern Ontario."

76. *Sault Star*, 1 December 1917; Heron, "Working-Class Hamilton," pp. 354-55, 358; Hamilton *Spectator*, 29 January, 8, 14 February 1919; PAO, RG 23, E-30, 1.6; *LN*, 30 September 1920. See also David Montgomery, "Immigrants, Industrial Unions, and Social Reconstruction in the United States, 1916-1923," *L/LT*, 13 (Spring, 1984), pp. 111-13.

77. One of Stelco's spies reported in February, 1919, "The aliens are very much afraid that if they do not keep quiet, they will be deported and the fact that such action will be taken, is having a great effect upon them." PAO, RG 23, E-30, 1.6. A police spy reached the same conclusion a year later. PAC, MG 27, II, D 19 (A.L. Sifton Papers), Vol. 9. On the actual policies of repression against these European radicals, first through the wartime measures passed in the fall of 1918 and then through Section 41 of the Immigration Act (against "anarchists and Bolsheviks"), see Avery, *"Dangerous Foreigners"*, pp. 82-89; Ian Angus, *Cana-*

dian Bolsheviks: The Early Years of the Communist Party of Canada (Montreal, 1981), pp. 39-42; Heron, "Working-Class Hamilton," pp. 355-59; Halifax *Herald*, 17 June 1919.

78. See, for example, *AJ*, 19 February 1920; *MLH*, 20 January 1923.
79. See Gary S. Cross, "The Quest for Leisure: Reassessing the Eight-Hour Day in France," *Journal of Social History* (Winter, 1984), pp. 195-216.
80. RCIRE, New Glasgow, pp. 3538, 3816, 3866; Halifax *Herald*, 22 February, 21 March, 15, 21 April, 21 May, 13, 27 June 1919; PAC, RG 27, Vol. 317, File 19(330); Vol. 324, File 20(385); *WW*, 26 March, 9, 16, 23 April 1920.
81. *EF*, 19, 26 April, 24 May, 7 June 1919; Halifax *Herald*, 6, 7 March, 15, 21 April, 19, 20 May, 5 June 1919.
82. *AJ*, 26 June 1919; *Sault Star*, 16 June 1919.
83. National Industrial Conference, *Proceedings*, pp. 52-98.
84. *IC*, 20,7 (July, 1919), pp. 209-10; 20,10 (October, 1919), p. 50; *LG*, 25,8 (August, 1925), pp. 784-85; 27,1 (January, 1927), p. 102; 29,6 (June, 1929), p. 628; Hucul, "Canada and the International Labour Organization," pp. 111-279.
85. MacEwan, *Miners and Steelworkers*, pp. 71-72; PANS, MG 2, Vol. 38, F4/10978-80 (R.M. Wolvin to E.A. Armstrong, 2, 13, 14 April 1924); Vol. 676, F6/15734 (R.M. Wolvin to E.A. Armstrong, 2 April 1924); *MLH*, 29 April, 6, 13 May 1922; David Hoffman, "Intra-Party Democracy: A Case Study," *Canadian Journal of Economics and Political Science*, 27,2 (May, 1961), pp. 223-35.
86. RCIRE, Sydney, p. 3811.
87. Brody, *Labor in Crisis*.
88. Kealey, "1919: The Canadian Labour Revolt."
89. The Canadian steelmaking corporations had little experience of working closely together on any matters, but in January, 1919, a newspaperman noticed that several corporate leaders in the industry had checked in to the King Edward Hotel in Toronto, undoubtedly for some private strategizing. *Sault Star*, 11 January 1919.
90. The need for more careful, concerted action in relation to labour was reflected in the decision of the Canadian Manufacturers' Association to create an Industrial Relations Committee, which included two Stelco executives, and in the work of the pro-business Canadian Reconstruction Association in reaching out to "moderate" labour leaders. *IC*, August, 1919; Tom Traves, *State and Enterprise*, pp. 24-27.
91. See, for example, Disco's statement in Halifax *Herald*, 1 January 1919.
92. Avery, *"Dangerous Foreigners"*, pp. 82-89; Elliott Samuels, "The Red Scare in Ontario: The Reaction of the Ontario Press to the Internal and External Threat of Bolshevism, 1917-1919" (M.A. thesis, Queen's University, 1971).
93. *ISC*, 3,1 (February, 1920), p. 34; *LN*, 5 March 1920; *ND*, 18 March 1920; Halifax *Herald*, 13 April 1920; *WW*, 25 June 1920.
94. PAC, RG 27, Vol. 320, File 20(125); *LN*, 7 May 1920; *ND*, 20 May 1920.
95. *AJ*, 22, 29 July, 5, 26 August, 2, 16, 23, 30 September, 2 December 1920.
96. Heron and De Zwaan, "Industrial Unionism in Eastern Ontario"; Montgomery, "Stelco Story," p. 5; Montgomery interview.
97. PAC, RG 27, Vol. 320, File 20(107).

98. Halifax *Herald*, 9 August 1920.

99. *LG*, 21,1 (January, 1921), p. 41; 21,12 (December, 1921), p. 1479; Halifax *Herald*, 23, 24, 26 November, 10, 11, 18, 23 December, 4, 8, 18 January 1921; *Canadian Railroader*, 5 February 1921; PAC, RG 27, Vol. 326, File 21(158); *ISC*, 3,11 (December, 1920), p. 323.

100. PAC, RG 27, Vol. 324, File 20(385); *Sault Star*, 18-20, 22, 24, 27, 29, 30 November 1920; AUCAI.

101. The reports of the Canadian lodges to the *Amalgamated Journal* from late 1920 through 1922, when most of them succumbed, were full of these stories of layoffs, plant closings, and short time.

102. *ISC*, 3,12 (January, 1921), p. 365; 4,1 (February, 1921), p. 9; 4,5 (June, 1921), p. 142; *Sault Star*, 14 January 1921; *AJ*, 20 January, 17 February 1921; *LG*, 21,2 (February, 1921), p. 147; 21,3 (March, 1921), p. 314; *MLH*, 6 May 1922.

103. Curtis, "Akes and Pains," p. 6; AUCAI.

104. Brody, *Labor in Crisis*, pp. 166-69.

105. *ND*, 24 June, 15, 29 July 1920; *LN*, 25 June 1920; *AJ*, 15 July 1920; Montgomery, "Stelco Story."

106. *Sault Star*, 1 September, 24, 27 November 1920; *AJ*, 11, 18 November 1920. The Sault's Amalgamated leaders, and the union's Canadian vice-president, Ernest Curtis, looked for support at the 1921 convention of the Trades and Labor Congress of Canada, but their resolution in favour of industrial unionism was soundly defeated. Trades and Labor Congress of Canada, *Proceedings*, 1921, p. 211; AUCAI; Peter Warrian interview with Ted Barbet.

107. *AJ*, 28 September 1922, 7, 14, 21 February 1924; see also Bishop's *Recollections of the "Amalgamated"* (n.p., n.d.), pp. 3, 5.

108. *LG*, 24,7 (July, 1924), p. 569; Canada, Department of Labour, *Labour Organization in Canada* (Ottawa, 1924), p. 206.

109. Curtis, "Akes and Pains," p. 6; *MLH*, 4, 18 November 1922. These unions dissolved back into the Amalgamated Association when organizing began again in the fall and winter of 1922.

110. David Frank, "Class Conflict in the Coal Industry, Cape Breton, 1922," in Kealey and Warrian, *Essays*, pp. 161-84; MacEwan, *Miners and Steelworkers*, pp. 79-91; Donald Macgillivray, "Industrial Unrest in Cape Breton, 1919-1925" (M.A. thesis, University of New Brunswick, 1971), Chapter II.

111. *MLH*, 4, 11, 18 November, 2, 30 December 1922; *AJ*, 28 September, 9, 23 November 1922; *Worker* (Toronto), 1 December 1922; BI, William G. Snow, "Sydney Steelworkers: Their Troubled Past and the Birth of Lodge 1064" (typescript, 1979), pp. 7-8; Macgillivray, "Industrial Unrest," p. 94. Disco released figures indicating that the plant council plan had been rejected by a margin of 1,562 to 1,021, but the union claimed that no more than 200 had voted and that many union men had refused to vote.

112. PAC, RG 27, Vol. 330, File 23(8); PANS, F7/12806, F7/12808; *MLH*, 3, 10, 17 February 1923; *AJ*, 15 February 1923; *WW*, 16 February, 2 March 1923; Canada, Commission to Inquire into the Industrial Unrest among the Steel Workers at Sydney, N.S., *Report* (Ottawa, 1924) (hereafter Robertson Commission), pp. 9-12; Snow, "Sydney Steelworkers," pp. 8-9; Curtis, "Akes and

Pains," p. 7; BI, Tape 52 (Dan MacKay and Doane Curtis), Tape 158 (George MacEachern); MacEachern, "Autobiography," p. 44; Macgillivray, "Industrial Unrest," pp. 94-95; MacEwan, *Miners and Steelworkers*, p. 91.

113. *WW*, 6 April 1923. The charges would eventually be dismissed in June for lack of evidence, after considerable fund-raising to raise a defence fund for the men. *MLH*, 23, 30 June 1923.

114. PANS, MG 2, Vol. 666 (E.H. Armstrong Papers), F7/12826 (P.McNeil to E.H. Armstrong, 10 April 1923).

115. *MLH*, 14 April 1923; PANS, MG 2, Vol. 666, F7/12809, F7/12817, F7/12819, F7/12826, F7/12829, F7/12832, F7/12839a, F8/12844A, F8/12857, F8/12863, Vol. 670, F5/14018, F5/14019, F5/14020, F5/14023, F5/14025; Macgillivray, "Industrial Unrest," pp. 95-107; David Frank, "The Trial of J.B. McLachlan," Canadian Historical Association, *Historical Papers*, 1983, pp. 210-11; Snow, "Sydney Steelworkers," pp. 9-10; Curtis, "Akes and Pains," pp. 7-8; MacEwan, *Miners and Steelworkers*, pp. 92-94.

116. *MLH*, 3, 10, 31 March, 12, 26 May, 2, 9, 16, 30 June 1923; PANS, MG 2, Vol. 666, F8/12841; Macgillivray, "Industrial Unrest," p. 108; Snow, "Sydney Steelworkers," pp. 10-11; Robertson Commission, pp. 12-13.

117. This advice was Tighe's only involvement in the strike, and because the Sydney workers had not followed the Amalgamated's formal procedures for calling a strike, he would not authorize strike pay from international headquarters. See Robertson's interview with him in University of British Columbia Library, Special Collections, James Robertson Papers, Box 5, File 1, "At Pittsburgh, Dec. 18th/23."

118. "The 1923 Strike in Steel and the Miners' Sympathy Strike," pp. 1-9; Macgillivray, "Industrial Unrest," pp. 108-14; Macgillivray, "Military Aid to the Civil Power: The Cape Breton Experience in the 1920's," in Don Macgillivray and Brian Tennyson, eds., *Cape Breton Historical Essays* (Sydney, 1980), pp. 102-03; *MLH*, 7 July 1923; *Worker* (Toronto), 1 August 1923; Snow, "Sydney Steelworkers," pp. 11-13; MacEwan, *Miners and Steelworkers*, pp. 95-97; Robertson Commission, pp. 13-16; "Reminiscences of a Sydney Steelworker: Frank Smith - Local 1064 and What It Achieved," *New Maritimes*, 4,7 (March, 1986), pp. 4-5.

119. "1923 Strike in Steel," pp. 9-13; Frank, "Trial," pp. 211-12; Macgillivray, "Military Aid," pp. 103-04; MacEwan, *Miners and Steelworkers*, pp. 97-98; PANS, MG 2, Vol. 666, F8 and F9; *AJ*, 26 July 1923, 3 January 1924.

120. Frank, "Trial," pp. 213-25; Macgillivray, "Industrial Unrest," pp. 115-31; MacEwan, *Miners and Steelworkers*, pp. 98-109; Snow, "Sydney Steelworkers," pp. 13-16; "1923 Strike in Steel," pp. 13-15; *MLH*, 4 August 1923; PANS, MG 2, Vol. 666, F8/12857, F8/12860, F8/12863, F8/12871; Vol. 670, 13990. On Clarence MacKinnon, see "The Great Debate," *New Maritimes*, 3,7 (April, 1985), pp. 4-10.

121. "1923 Strike in Steel," p. 15; MacKay and Curtis interview; *MLH*, 11, 18 August 1923; Macgillivray, "Industrial Unrest," pp. 136-37; MacEwan, *Miners and Steelworkers*, pp. 109-10; PANS, MG 2, Vol. 670, F/14010. The provincial police commissioner estimated that "considerably more than 500" were not rehired. The radical Ukrainian community, which had been prominent in orga-

nizing the European workers, lost more than a hundred men who were forced to migrate.

122. PANS, MG 2, Vol. 666, F8/12873, F8/12875, F8/12877, F8/12884; Macgillivray, "Industrial Unrest," p. 136; Robertson Commission, pp. 17-19.

123. Robertson Commission; Macgillivray, "Industrial Unrest," pp. 137, 143-48; Macgillivray, "Military Aid," pp. 104-06; L.W. Bentley, "Aid to the Civil Power: Social and Political Aspects, 1904-1924," *Canadian Defence Quarterly*, 8,1 (Summer, 1978), pp. 44-51.

124. Sydney *Post*, 18 December 1923; Curtis, "Akes and Pains," pp. 9-15; BI, Tape 532 (Carl Neville); MacEachern interview; Robertson Commission, p. 18. The One Big Union was active in several Nova Scotia industrial centres, especially mining towns, between 1924 and 1926, but was eventually driven out by the local UMW, including those with Communist sympathies; see E.J. Shields, "A History of Trade Unionism in Nova Scotia" (M.A. thesis, Dalhousie University, 1945), pp. 58-64; James M. Cameron, *The Pictonian Colliers* (Halifax, 1974), pp. 152-54; David J. Bercuson, *Fools and Wise Men: The Rise and Fall of the One Big Union* (Toronto, 1978), pp. 238-45. The decision to leave the Amalgamated Association late in 1923 was not surprising since even before the strike the local leadership had publicly denounced the international's "reactionary officers." *MLH*, 28 April 1923; *WW*, 22 June 1923.

125. The provincial government's willingness to support Disco in breaking the Sydney steelworkers' union and attacking radicals deserves a full study in itself. Its prime concern was in protecting the all-important revenues from coal royalties. At the same time, in this pre-World War II period Canadian provincial governments were also so keenly anxious to maintain a good investment climate within their jurisdictions that they feared the spectre of militant unionism and radicalism might threaten the attractions of cheap and docile labour.

Conclusion

1. Paul Craven and Tom Traves, "Dimensions of Paternalism: Discipline and Culture in Canadian Railway Operations in the 1850s," in Heron and Storey, eds., *On the Job: Confronting the Labour Process in Canada*, pp. 47-74; Craven and Traves, "Labour and Management in Canadian Railway Operations: The First Decade," paper presented to the Commonwealth Labour History Conference, Coventry, England, 1981; Walter Licht, *Working for the Railroad: The Organization of Work in the Nineteenth Century* (Princeton, N.J., 1983); Desmond Morton, "Taking on the Grand Trunk: The Locomotive Engineers Strike of 1876-7," *L/LT*, 2 (1977), pp. 5-34; Joseph Hugh Tuck, "The Canadian Railways and the International Brotherhoods: Labour Organization in the Railway Running Trades in Canada, 1865-1914" (Ph.D. thesis, University of Western Ontario, 1975).

2. See Traves, *State and Enterprise*; William L. Marr and Donald G. Paterson, *Canada: An Economic History* (Toronto, 1980), pp. 375-417.

3. See, for example, Montgomery, *Workers' Control in America*, p. 10.

4. Piore, *Birds of Passage*; Stephen Castles and Godula Kosack, *Immigrant Workers and Class Structures in Western Europe* (London, 1973); Dirk Hoerder,

"Immigration and the Working Class: The Remigration Factor," *International Labor and Working Class History*, 21 (Spring, 1982), pp. 28-41.

5. Meyer, *The Five Dollar Day*, pp. 75-77; Joyce Shaw Peterson, "Auto Workers and Their Work, 1900-1933," *Labor History*, 22,2 (Spring, 1981), pp. 223-26; John Manley, "Organize the Unorganized: Communists and the Struggle for Industrial Unionism in the Canadian Automobile Industry," paper presented to the Canadian Historical Association, 1981, p. 6; James R. Barrett, "Unity and Fragmentation: Class, Race, and Ethnicity on Chicago's South Side, 1900-1922," *Journal of Social History* (Fall, 1984), pp. 37-55; Robert Ozanne, *A Century of Labor-Management Relations at McCormick and International Harvester* (Madison, 1967), p. xvi; Heron, "Working-Class Hamilton," pp. 310-11; Stanley Scott, "A Profusion of Issues: Immigrant Labour, the World War, and the Cominco Strike of 1917," *L/LT*, 2 (1977), pp. 54-78.

6. These conclusions parallel those of Jean Morrison, "Ethnicity and Violence: The Lakehead Freight Handlers Before World War I," in Kealey and Warrian, eds., *Essays in Canadian Working Class History*, pp. 143-60.

7. Gordon, Edwards, and Reich, *Segmented Work, Divided Workers*, pp. 100-64.

8. Storey, "Workers, Unions, and Steel"; Ian Radforth, "Bushworkers and Bosses: The Social History of the Logging Industry of Northern Ontario, 1900-1980" (Ph.D. thesis, York University, 1985).

9. William Lazonick, "Industrial Relations and Technical Change: The Case of the Self-Acting Mule," *Cambridge Journal of Economics*, 3 (1979); Jonathan Zeitlin, "Craft Control and the Division of Labour: Engineers and Compositors in Britain, 1890-1930," *ibid.*, pp. 263-74; Raphael Samuel, "Workshop of the World: Steam Power and Hand Technology in Mid-Victorian Britain," *History Workshop*, 3 (Spring, 1977), pp. 6-72.

10. On the importance of skilled paper makers in early twentieth-century pulp and paper production, see William E. Greening, *Paper Makers in Canada: A History of the Paper Makers' Union in Canada* (Cornwall, 1952); Gil Schonning, "Union-Management Relations in the Pulp and Paper Industry in Ontario and Quebec, 1914-1950" (Ph.D. thesis, University of Toronto, 1955). Gail Cuthbert-Brandt has also shown the continuing importance of skilled and semi-skilled workers in the Quebec cotton industry in this period; see "The Transformation of Women's Work in the Quebec Cotton Industry, 1920-1950," in Bryan D. Palmer, ed., *The Character of Class Struggle: Essays in Canadian Working-Class History* (Toronto, 1986), pp. 115-34.

11. Nelson Lichtenstein, "Auto Worker Militancy and the Structure of Factory Life, 1937-1955," *Journal of American History*, 67,2 (September, 1980), pp. 336-38.

12. Meyer, *The Five Dollar Day*, pp. 95-123.

13. Andrew L. Friedman, *Industry and Labour: Class Struggle at Work and Monopoly Capitalism* (London, 1977).

14. For the limits of Taylorism in the United States, see Bryan D. Palmer, "Class, Conception and Conflict: The Thrust for Efficiency, Managerial View of Labor and the Working Class Rebellion, 1903-1922," *Review of Radical Political Economics*, 7 (Summer, 1975), pp. 31-49; Nelson, *Managers and Workers*, pp. 55-78. For Britain, see Littler, *The Development of the Labour Process in Capitalist Societies*.

15. Lichtenstein, "Auto Worker Militancy"; Peter Friedlander, *The Emergence of a UAW Local, 1936-1939: A Study in Class and Culture* (Pittsburgh, 1975); John Manley, "Organize the Unorganized"; Schonning, "Union-Management Relations"; John Tait Montague, "Trade Unionism in the Canadian Meatpacking Industry" (Ph.D. thesis, University of Toronto, 1950); David Brody, *The Butcher Workmen: A Study of Unionization* (Cambridge, Mass., 1964); Schatz, *The Electrical Workers*. See also Jeffrey Miles Haydu, "Factory Politics in the British and American Metal Trades: Changing Agenda for Protest, 1890-1922" (Ph.D. thesis, University of California, Berkeley, 1984), for a valuable discussion of what he calls "selective mobilization."

16. See, for example, Manley, "Organize the Unorganized"; and Wayne Roberts, ed., *Organizing Westinghouse: Alf Ready's Story* (Hamilton, 1979); Ralph Ellis, "The Unionization of a Mill Town: Cornwall in 1936," *The Register*, 2,1 (March, 1981), pp. 83-101.

17. Meyer, *The Five Dollar Day*; Allan Nevins, *Ford: Decline and Rebirth* (New York, 1963), pp. 150-51.

18. Charles Sabel has pointed to the importance of the new ideological climate that helped shape Italy's "hot autumn" of 1969; see Sabel, *Work and Politics*, pp. 145-67.

19. John Bodnar, "Immigration, Kinship, and the Rise of Working-Class Realism in Industrial America," *Journal of American History*, 14,1 (Fall, 1980), pp. 45-65; Peterson, "Auto Workers," p. 236; Storey, "Workers, Unions, and Steel."

20. See, in particular, Hareven, *Family Time and Industrial Time*; Bodnar, "Immigration."

21. Storey, "Workers, Unions, and Steel," pp. 179-244; MacEachern, "Autobiography"; Waisglass, "Case Study"; Irving Martin Abella, *Nationalism, Communism, and Canadian Labour: The CIO, the Communist Party, and the Canadian Congress of Labour* (Toronto, 1973), pp. 55-60; Ferris, *Algoma's Industrial and Trade Union Development*, pp. 116-18; Ronald McDonald Adams, "The Development of the United Steel Workers of America in Canada, 1936-1951" (M.A. thesis, Queen's University, 1952).

22. Storey, "Workers, Unions, and Steel," pp. 297-418; MacDowell, "Formation of Canadian Industrial Relations System"; MacDowell, "The 1943 Steel Strike Against Wartime Wage Controls," *L/LT*, 10 (Autumn, 1982), pp. 65-85; Adams, "Development," pp. 129-56; Roberts, *Baptism*; C.D. Martin, "The 1946 Steel Strike," in *50 Years of Algoma Labour*, pp. 101-16.

23. Ronald B. Bean, "Joint Union-Management Job Evaluation in the Canadian Steel Industry," *Relations industrielles/Industrial Relations*, 17,1 (April, 1962), pp. 115-26; Bean, "The 'Co-operative Wage Study' and the Canadian Steelworkers," *ibid.*, 19,1 (January, 1964), pp. 55-70; C. Bryan Williams, "Collective Bargaining and Wage Equalization in Canada's Iron and Steel Industry, 1939-1964," *ibid.*, 26,2 (April, 1971), pp. 308-44.

24. Kilbourn, *History of the Steel Company of Canada*, pp. 199-200.

25. See, for example, Maxwell Flood, *Wildcat Strike in Lake City* (Ottawa, 1968), which is based on the Stelco wildcat of 1966; Bill Freeman, *1005: Political Life in a Union Local* (Toronto, 1982), pp. 99-160.

Index

THE CANADIAN SOCIAL HISTORY SERIES